CALCULUS OF FINITE DIFFERENCES

BY

GEORGE BOOLE, D.C.L.

EDITED BY

J. F. MOULTON

CHELSEA PUBLISHING COMPANY

NEW YORK, N.Y.

FIFTH EDITION

THE PRESENT, FIFTH EDITION DIFFERS FROM THE FOURTH EDITION (NEW YORK, 1958) PRIMARILY IN CERTAIN IMPROVEMENTS IN NOTATION AND THE ADDITION OF A NAME AND SUBJECT INDEX. THE THIRD, AND EARLIER EDITIONS WERE PUBLISHED UNDER THE TITLE: A TREATISE ON THE CALCULUS OF FINITE DIFFERENCES

PUBLISHED (ON ALKALINE PAPER) NEW YORK, N.Y., 1970

LIBRARY OF CONGRESS CATALOG CARD NUMBER 76-119364

INTERNATIONAL STANDARD BOOK NUMBER 0-8284-1121-2

PRINTED IN THE UNITED STATES OF AMERICA

PREFACE TO THE FIRST EDITION.

In the following exposition of the Calculus of Finite Differences, particular attention has been paid to the connexion of its methods with those of the Differential Calculus—a connexion which in some instances involves far more than a merely formal analogy.

Indeed the work is in some measure designed as a sequel to my *Treatise on Differential Equations*. And it has been composed on the same plan.

Mr Stirling, of Trinity College, Cambridge, has rendered me much valuable assistance in the revision of the proof-sheets. In offering him my best thanks for his kind aid, I am led to express a hope that the work will be found to be free from important errors.

GEORGE BOOLE.

Queen's College, Cork,
April 18, 1860.

PREFACE TO THE SECOND EDITION.

When I commenced to prepare for the press a Second Edition of the late Dr Boole's Treatise on Finite Differences, my intention was to leave the work unchanged save by the insertion of sundry additions in the shape of paragraphs marked off from the rest of the text. But I soon found that adherence to such a principle would greatly lessen the value of the book as a Text-book, since it would be impossible to avoid confused arrangement and even much repetition. I have therefore allowed myself considerable freedom as regards the form and arrangement of those parts where the additions are considerable, but I have strictly adhered to the principle of inserting all that was contained in the First Edition.

As such Treatises as the present are in close connexion with the course of Mathematical Study at the University of Cambridge, there is considerable difficulty in deciding the question how far they should aim at being exhaustive. I have held it best not to insert investigations that involve complicated analysis unless they possess great suggestiveness or are the bases of important developments of the subject. Under the present system the premium on wide superficial reading is so great that such investigations, if inserted, would seldom be read. But though this is at present the case, there is every reason to hope that it will not continue to be so; and in view of a time when students will aim at an exhaustive study of a few subjects in preference to a superficial acquaintance with the whole range of Mathematical research, I have added brief notes referring to most of the papers on the subjects of this Treatise that have appeared in the Mathematical Serials, and to other original sources. In virtue of such references, and the brief indication of the subject of the paper that accompanies each, it is hoped that this work may serve as a handbook to students who wish to read the subject more thoroughly than they could do by confining themselves to an Educational Text-book.

The latter part of the book has been left untouched. Much of it I hold to be unsuited to a work like the present, partly for reasons similar to those given above, and partly because it treats in a brief and necessarily imperfect manner subjects that had better be left to separate treatises. It is impossible within the limits of the present work to treat adequately the Calculus of Operations and the Calculus of Functions, and I should have preferred leaving them wholly to such treatises as those of Lagrange, Babbage, Carmichael, De Morgan, &c. I have therefore abstained from making any additions to these portions of the book, and have made it my chief aim to render more evident the remarkable analogy between the Calculus of Finite Differences and the Differential Calculus. With this view I have suffered myself to digress into the subject of the Singular Solutions of Differential Equations, to a much greater extent than Dr Boole had done. But I trust that the advantage of rendering the investigation a complete one will be held to justify the irrelevance of much of it to that which is nominally the subject of the book. It is partly from similar considerations that I have adopted a nomenclature slightly differing from that commonly used (e.g. Partial Difference-Equations for Equations of Partial Differences).

I am greatly indebted to Mr R. T. Wright of Christ's College for his kind assistance. He has revised the proofs for me, and throughout the work has given me valuable suggestions of which I have made free use.

JOHN F. MOULTON.

CHRIST'S COLLEGE,
Oct. 1872.

CONTENTS.

DIFFERENCE- AND SUM-CALCULUS.

CHAPTER IV.

CHAPTER V.

CHAPTER VI.

CHAPTER VII.

CHAPTER VIII.

DIFFERENCE- AND FUNCTIONAL EQUATIONS.

CHAPTER IX.

CHAPTER X.

CHAPTER XI.

CHAPTER XII.

CHAPTER XIII.

CHAPTER XIV.

CHAPTER XV.

CHAPTER XVI.

FINITE DIFFERENCES.

CHAPTER I.

NATURE OF THE CALCULUS OF FINITE DIFFERENCES.

1. THE Calculus of Finite Differences may be strictly defined as the science which is occupied about the ratios of the simultaneous increments of quantities mutually dependent. The Differential Calculus is occupied about the *limits* to which such ratios approach as the increments are indefinitely diminished.

In the latter branch of analysis if we represent the independent variable by x, any dependent variable considered as a function of x is represented primarily indeed by $\phi(x)$, but, when the rules of differentiation founded on its functional character are established, by a single letter, as u. In the notation of the Calculus of Finite Differences these modes of expression seem to be in some measure blended. The dependent function of x is represented by u_x, the suffix taking the place of the symbol which in the former mode of notation is enclosed in brackets. Thus, if $u_x \equiv \phi(x)$, then

$$u_{x+h} = \phi(x+h),$$

$$u_{\sin x} = \phi(\sin x),$$

and so on. But this mode of expression rests only on a convention, and as it was adopted for convenience, so when convenience demands it is laid aside.

The step of transition from a function of x to its increment, and still further to the *ratio* which that increment bears to the increment of x, may be contemplated apart from its subject, and it is often important that it should be so contemplated, as an operation governed by laws. Let then Δ, prefixed to the expression of any function of x, denote the operation of taking the increment of that function corresponding to a given constant increment Δx of the variable x. Then, representing as above the proposed function of x by u_x, we have

$$\Delta u_x = u_{x+\Delta x} - u_x,$$

and

$$\frac{\Delta u_x}{\Delta x} = \frac{u_{x+\Delta x} - u_x}{\Delta x}.$$

Here then we might say that as $\frac{d}{dx}$ is the fundamental operation of the Differential Calculus, so $\frac{\Delta}{\Delta x}$ is the fundamental operation of the Calculus of Finite Differences.

But there is a difference between the two cases which ought to be noted. In the Differential Calculus $\frac{du}{dx}$ is not a true fraction, nor have du and dx any distinct meaning as symbols of quantity. The fractional form is adopted to express the limit to which a true fraction approaches. Hence $\frac{d}{dx}$, and not d, there represents a real operation. But in the Calculus of Finite Differences $\frac{\Delta u_x}{\Delta x}$ is a true fraction. Its numerator Δu_x stands for an actual magnitude. Hence Δ might itself be taken as the fundamental operation of this Calculus, always supposing the actual value of Δx to be given; and the Calculus of Finite Differences might, in its symbolical character, be defined either as the science of the laws of the operation Δ, the value of Δx being supposed given, or as the science of the laws of the operation $\frac{\Delta}{\Delta x}$. In consequence of the fundamental difference above noted between the Differential Calculus and the Calculus of Finite Differences, the term Finite ceases to be necessary as a mark of distinction. The former is a calculus of *limits*, not of *differences*.

2. Though Δx admits of any constant value, the value usually given to it is unity. There are two reasons for this.

First. The Calculus of Finite Differences has for its chief subject of application the terms of series. Now the law of a series, however expressed, has for its ultimate object the determination of the values of the successive terms as dependent upon their numerical order and position. Explicitly or implicitly, each term is a *function* of the integer which expresses its position in the series. And thus, to revert to language familiar in the Differential Calculus, the independent variable admits only of integral values whose common difference is unity. For instance, in the series of terms

$$1^2,\ 2^2,\ 3^2,\ 4^2,\ \ldots$$

the general or x^{th} term is x^2. It is an explicit function of x, but the values of x are the series of natural numbers, and $\Delta x = 1$.

Secondly. When the general term of a series is a function of an independent variable t whose successive differences are constant but not equal to unity, it is always possible to replace that independent variable by another, x, whose common difference shall be unity. Let $\phi(t)$ be the general term of the series, and let $\Delta t = h$; then assuming $t = hx$ we have $\Delta t = h\Delta x$, whence $\Delta x = 1$.

Thus it suffices to establish the rules of the Calculus on the assumption that the finite difference of the independent variable is unity. At the same time it will be noted that this assumption reduces to equivalence the symbols $\frac{\Delta}{\Delta x}$ and Δ.

We shall therefore in the following chapters develope the theory of the operation denoted by Δ and defined by the equation

$$\Delta u_x = u_{x+1} - u_x.$$

But we shall, where convenience suggests, consider the more general operation

$$\frac{\Delta u_x}{\Delta x} = \frac{u_{x+h} - u_x}{h},$$

where $\Delta x = h$.

CHAPTER II.

DIRECT THEOREMS OF FINITE DIFFERENCES.

1. THE operation denoted by Δ is capable of repetition. For the difference of a function of x, being itself a function of x, is subject to operations of the same kind.

In accordance with the algebraic notation of indices, the difference of the difference of a function of x, usually called the second difference, is expressed by attaching the index 2 to the symbol Δ. Thus

$$\Delta\Delta u_x \equiv \Delta^2 u_x.$$

In like manner

$$\Delta\Delta^2 u_x \equiv \Delta^3 u_x,$$

and generally

$$\Delta\Delta^{n-1} u_x \equiv \Delta^n u_x \qquad (1),$$

the last member being termed the nth difference of the function u_x. If we suppose $u_x \equiv x^3$, the successive values of u_x with their successive differences of the first, second, and third orders will be represented in the following scheme:

Values of x	1	2	3	4	5	6 ...
u_x	1	8	27	64	125	216 ...
Δu_x	7	19	37	61	91 ...	
$\Delta^2 u_x$	12	18	24	30 ...		
$\Delta^3 u_x$	6	6	6 ...			

It may be observed that each set of differences may either be formed from the preceding set by successive subtractions in accordance with the definition of the symbol Δ, or calculated from the general expressions for Δu, $\Delta^2 u$, ... by assigning to x the successive values 1, 2, 3, ... Since $u_x = x^3$, we shall have

$$\Delta u_x = (x+1)^3 - x^3 = 3x^2 + 3x + 1,$$

$$\Delta^2 u_x = \Delta (3x^2 + 3x + 1) = 6x + 6,$$

$$\Delta^3 u_x = 6.$$

It may also be noted that the third differences are here constant. And generally *if u_x be a rational and integral function of x of the nth degree, its nth differences will be constant.* For let

$$u_x \equiv ax^n + bx^{n-1} + \ldots,$$

then

$$\Delta u_x = a(x+1)^n + b(x+1)^{n-1} + \ldots$$
$$- ax^n - bx^{n-1} - \ldots$$
$$= anx^{n-1} + b_1 x^{n-2} + b_2 x^{n-3} + \ldots,$$

b_1, b_2, ... being constant coefficients. Hence Δu_x is a rational and integral function of x of the degree $n-1$. Repeating the process, we have

$$\Delta^2 u_x = an(n-1)x^{n-2} + c_1 x^{n-3} + c_2 x^{n-4} + \ldots,$$

a rational and integral function of the degree $n-2$; and so on.

Finally we shall have

$$\Delta^n u_x = an(n-1)(n-2)\ldots 1,$$

a constant quantity.

Hence also we have

$$\Delta^n x^n = 1 \,.\, 2 \ldots n \qquad (2).$$

2. While the operation or series of operations denoted by $\Delta, \Delta^2, \ldots, \Delta^n$ are always possible when the subject-function u_x is given, there are certain elementary cases in which the forms of the results are deserving of particular attention, and these we shall next consider.

Differences of Elementary Functions.

1st. Let $u_x \equiv x(x-1)(x-2)\ldots(x-m+1)$.

Then by definition,

$$\Delta u_x = (x+1)x(x-1)\ldots(x-m+2) - x(x-1)(x-2)\ldots(x-m+1)$$
$$= mx(x-1)(x-2)\ldots(x-m+2).$$

When the factors of a continued product increase or decrease by a constant difference, or when they are similar functions of a variable which, in passing from one to the other, increases or decreases by a constant difference, as in the expression

$$\sin x \sin(x+h)\sin(x+2h)\ldots\sin\{x+(m-1)h\},$$

the factors are usually called *factorials*, and the term in which they are involved is called a factorial term. For the particular kind of factorials illustrated in the above example it is common to employ the notation

$$x(x-1)\ldots(x-m+1) \equiv x^{(m)} \qquad (1),$$

doing which, we have

$$\Delta x^{(m)} = mx^{(m-1)} \qquad (2).$$

Hence, $x^{(m-1)}$ being also a factorial term,

$$\Delta^2 x^{(m)} = m(m-1)x^{(m-2)},$$

and generally

$$\Delta^n x^{(m)} = m(m-1)\ldots(m-n+1)x^{(m-n)} \qquad (3).$$

2ndly. Let $u_x \equiv \dfrac{1}{x(x+1)\ldots(x+m-1)}$.

Then by definition,

$$\Delta u_x = \frac{1}{(x+1)(x+2)\ldots(x+m)} - \frac{1}{x(x+1)\ldots(x+m-1)}$$
$$= \left(\frac{1}{x+m} - \frac{1}{x}\right)\frac{1}{(x+1)(x+2)\ldots(x+m-1)}$$
$$= \frac{-m}{x(x+1)\ldots(x+m)} \qquad (4).$$

Hence, adopting the notation

$$\frac{1}{x(x+1)\dots(x+m-1)} = x^{(-m)},$$

we have

$$\Delta x^{(-m)} = -mx^{(-m-1)} \qquad (5).$$

Hence by successive repetitions of the operation Δ,

$$\Delta^n x^{(-m)} = -m(-m-1)\dots(-m-n+1)\,x^{(-m-n)}$$

$$= (-1)^n m(m+1)\dots(m+n-1)\,x^{(-m-n)} \qquad (6),$$

and this may be regarded as an extension of (3).

3rdly. Employing the most general form of factorials, we find

$$\Delta u_x u_{x-1} \dots u_{x-m+1} = (u_{x+1} - u_{x-m+1}) \times u_x u_{x-1} \dots u_{x-m+2} \qquad (7),$$

$$\Delta \frac{1}{u_x u_{x+1} \dots u_{x+m-1}} = \frac{u_x - u_{x+m}}{u_x u_{x+1} \dots u_{x+m}} \qquad (8),$$

and in particular if $u_x = ax + b$,

$$\Delta u_x u_{x-1} \dots u_{x-m+1} = amu_x u_{x-1} \dots u_{x-m+2} \qquad (9),$$

$$\Delta \frac{1}{u_x u_{x+1} \dots u_{x+m-1}} = \frac{-am}{u_x u_{x+1} \dots u_{x+m}} \qquad (10).$$

In like manner we have

$$\Delta \log u_x = \log u_{x+1} - \log u_x = \log \frac{u_{x+1}}{u_x}.$$

To this result we may give the form

$$\Delta \log u_x = \log\left(1 + \frac{\Delta u_x}{u_x}\right) \qquad (11).$$

So also

$$\Delta \log (u_x u_{x-1} \dots u_{x-m+1}) = \log \frac{u_{x+1}}{u_{x-m+1}} \qquad (12).$$

4thly. To find the successive differences of a^x.

We have

$$\Delta a^x = a^{x+1} - a^x$$
$$= (a-1)\, a^x \qquad (13).$$

Hence

$$\Delta^2 a^x = (a-1)^2\, a^x,$$

and generally,

$$\Delta^n a^x = (a-1)^n\, a^x \qquad (14).$$

Hence also, since $a^{mx} = (a^m)^x$, we have

$$\Delta^n a^{mx} = (a^m - 1)^n\, a^{mx} \qquad (15).$$

5thly. To deduce the successive differences of $\sin(ax+b)$ and $\cos(ax+b)$.

$$\Delta \sin(ax+b) = \sin(ax+b+a) - \sin(ax+b)$$
$$= 2\sin\frac{a}{2}\cos\left(ax+b+\frac{a}{2}\right)$$
$$= 2\sin\frac{a}{2}\sin\left(ax+b+\frac{a+\pi}{2}\right).$$

By inspection of the form of this result we see that

$$\Delta^2 \sin(ax+b) = \left(2\sin\frac{a}{2}\right)^2 \sin(ax+b+a+\pi) \qquad (16).$$

And generally,

$$\Delta^n \sin(ax+b) = \left(2\sin\frac{a}{2}\right)^n \sin\left\{ax+b+\frac{n(a+\pi)}{2}\right\} \qquad (17).$$

In the same way it will be found that

$$\Delta^n \cos(ax+b) = \left(2\sin\frac{a}{2}\right)^n \cos\left\{ax+b+\frac{n(a+\pi)}{2}\right\} \qquad (18).$$

These results might also be deduced by substituting for the sines and cosines their exponential values and applying (15).

3. The above are the most important forms. The following are added merely for the sake of exercise.

To find the differences of $\tan u_x$ and of $\tan^{-1}u_x$.

$$\begin{aligned}\Delta \tan u_x &= \tan u_{x+1} - \tan u_x \\ &= \frac{\sin u_{x+1}}{\cos u_{x+1}} - \frac{\sin u_x}{\cos u_x} \\ &= \frac{\sin (u_{x+1} - u_x)}{\cos u_{x+1} \cos u_x} \\ &= \frac{\sin \Delta u_x}{\cos u_{x+1} \cos u_x} \qquad (1).\end{aligned}$$

Next,

$$\begin{aligned}\Delta \tan^{-1} u_x &= \tan^{-1} u_{x+1} - \tan^{-1} u_x \\ &= \tan^{-1} \frac{u_{x+1} - u_x}{1 + u_{x+1} u_x} \\ &= \tan^{-1} \frac{\Delta u_x}{1 + u_{x+1} u_x} \qquad (2).\end{aligned}$$

From the above, or independently, it is easily shewn that

$$\Delta \tan ax = \frac{\sin a}{\cos ax \cos a(x+1)} \qquad (3),$$

$$\Delta \tan^{-1} ax = \tan^{-1} \frac{a}{1 + a^2 x + a^2 x^2} \qquad (4).$$

Additional examples will be found in the exercises at the end of this chapter.

4. When the increment of x is indeterminate, the operation denoted by $\frac{\Delta}{\Delta x}$ merges, on supposing Δx to become infinitesimal but the subject-function to remain unchanged, into the operation denoted by $\frac{d}{dx}$. The following are illustrations of the mode in which some of the general theorems of the Calculus of Finite Differences thus merge into theorems of the Differential Calculus.

Ex. We have

$$\frac{\Delta \sin x}{\Delta x} = \frac{\sin (x + \Delta x) - \sin x}{\Delta x}$$

$$= \frac{2 \sin \frac{1}{2} \Delta x \sin \left(x + \frac{\Delta x + \pi}{2}\right)}{\Delta x}.$$

And, repeating the operation n times,

$$\frac{\Delta^n \sin x}{(\Delta x)^n} = \frac{(2 \sin \frac{1}{2} \Delta x)^n \sin \left(x + n \frac{\Delta x + \pi}{2}\right)}{(\Delta x)^n} \qquad (1).$$

It is easy to see that the limiting form of this equation is

$$\frac{d^n \sin x}{dx^n} = \sin \left(x + \frac{n\pi}{2}\right) \qquad (2),$$

a known theorem of the Differential Calculus.

Again, we have

$$\frac{\Delta a^x}{\Delta x} = \frac{a^{x+\Delta x} - a^x}{\Delta x}$$

$$= \left(\frac{a^{\Delta x} - 1}{\Delta x}\right) a^x.$$

And hence, generally,

$$\frac{\Delta^n a^x}{(\Delta x)^n} = \left(\frac{a^{\Delta x} - 1}{\Delta x}\right)^n a^x \qquad (3).$$

Supposing Δx to become infinitesimal, this gives by the ordinary rule for vanishing fractions

$$\frac{d^n a^x}{dx^n} = (\log a)^n a^x \qquad (4).$$

But it is not from examples like these to be inferred that the Differential Calculus is merely a particular case of the Calculus of Finite Differences. The true nature of their connexion will be developed in a future chapter.

Expansion by factorials.

5. Attention has been directed to the formal analogy between the differences of factorials and the differential coefficients of powers. This analogy is further developed in the following proposition.

To develope $\phi(x)$, a given rational and integral function of x of the mth degree, in a series of factorials.

Assume

$$\phi(x) = a + bx + cx^{(2)} + dx^{(3)} \ldots + hx^{(m)} \qquad (1).$$

The legitimacy of this assumption is evident, for the new form represents a rational and integral function of x of the mth degree, containing a number of arbitrary coefficients equal to the number of coefficients in $\phi(x)$. And the actual values of the former might be determined by expressing both members of the equation in ascending powers of x, equating coefficients, and solving the linear equations which result. Instead of doing this, let us take the successive differences of (1). We find by (2), Art. 2,

$$\Delta\phi(x) = b + 2cx + 3dx^{(2)} \ldots + mhx^{(m-1)} \qquad (2),$$

$$\Delta^2\phi(x) = 2c + 3 \,.\, 2dx \ldots + m(m-1)\, hx^{(m-2)} \qquad (3),$$

....

$$\Delta^m\phi(x) = m(m-1) \ldots 1h \qquad (4).$$

And now making $x = 0$ in the series of equations (1)...(4), and representing by $\Delta\phi(0)$, $\Delta^2\phi(0)$, ... what $\Delta\phi(x)$, $\Delta^2\phi(x)$, ... become when $x = 0$, we have

$$\phi(0) = a, \quad \Delta\phi(0) = b, \quad \Delta^2\phi(0) = 2c,$$

....

$$\Delta^m\phi(0) = 1 \,.\, 2 \ldots mh.$$

Whence determining $a, b, c, \ldots h$, we have

$$\phi(x) = \phi(0) + \Delta\phi(0)\, x + \frac{\Delta^2\phi(0)}{2} x^{(2)} + \frac{\Delta^3\phi(0)}{2 \,.\, 3} x^{(3)} + \ldots \qquad (5).$$

If with greater generality we assume

$$\phi(x) = a + bx + cx(x-h) + dx(x-h)(x-2h) + \ldots,$$

we shall find by proceeding as before, (except in the employing of $\frac{\Delta}{\Delta x}$ for Δ, where $\Delta x = h$,)

$$\phi(x) = \{\phi(x)\} + \left\{\frac{\Delta\phi(x)}{\Delta x}\right\} x + \left\{\frac{\Delta^2\phi(x)}{(\Delta x)^2}\right\} \frac{x(x-h)}{1.2}$$
$$+ \left\{\frac{\Delta^3\phi(x)}{(\Delta x)^3}\right\} \frac{x(x-h)(x-2h)}{1.2.3} + \dots \quad (6),$$

where the brackets $\{\}$ denote that in the enclosed function, after reduction, x is to be made equal to 0.

Maclaurin's theorem is the limiting form to which the above theorem approaches when the increment Δx is indefinitely diminished.

General theorems expressing relations between the successive values, successive differences, and successive differential coefficients of functions.

6. In the equation of definition

$$\Delta u_x = u_{x+1} - u_x$$

we have the fundamental relation connecting the first difference of a function with two successive values of that function. Taylor's theorem gives us, if h be put equal to unity,

$$u_{x+1} - u_x = \frac{du_x}{dx} + \frac{1}{2}\frac{d^2u_x}{dx^2} + \frac{1}{2.3}\frac{d^3u_x}{dx^3} + \dots,$$

which is the fundamental relation connecting the first difference of a function with its successive differential coefficients. From these fundamental relations spring many general theorems expressing derived relations between the differences of the higher orders, the successive values, and the differential coefficients of functions.

As concerns the history of such theorems it may be observed that they appear to have been first suggested by particular instances, and then established, either by that kind of proof which consists in shewing that if a theorem is true for any particular integer value of an index n, it is true for the next greater value, and therefore for all succeeding values; or else by a peculiar method, hereafter to be explained, called the method of Generating Functions. But having been once established, the very forms of the theorems led to a deeper conception of their real nature, and it came to be understood that they were consequences of the formal laws of combination of those operations by which from a given function its succeeding values, its differences, and its differential coefficients are derived.

7. These progressive methods will be illustrated in the following example.

Ex. Required to express u_{x+n} in terms of u_x and its successive differences.

We have

$$u_{x+1} = u_x + \Delta u_x;$$

$$\therefore\ u_{x+2} = u_x + \Delta u_x + \Delta (u_x + \Delta u_x)$$

$$= u_x + 2\Delta u_x + \Delta^2 u_x.$$

Hence proceeding as before we find

$$u_{x+3} = u_x + 3\Delta u_x + 3\Delta^2 u_x + \Delta^3 u_x.$$

These special results suggest, by the agreement of their coefficients with those of the successive powers of a binomial, the general theorem

$$u_{x+n} = u_x + n\Delta u_x + \frac{n(n-1)}{1.2}\Delta^2 u_x + \frac{n(n-1)(n-2)}{1.2.3}\Delta^3 u_x + \ldots \quad (1).$$

Suppose then this theorem true for a particular value of n, then for the next greater value we have

$$u_{x+n+1} = u_x + n\Delta u_x + \frac{n(n-1)}{1.2}\Delta^2 u_x + \frac{n(n-1)(n-2)}{1.2.3}\Delta^3 u_x + \ldots$$

$$+ \Delta u_x + n\Delta^2 u_x + \frac{n(n-1)}{1.2}\Delta^3 u_x + \ldots$$

$$= u_x + (n+1)\Delta u_x + \frac{(n+1)n}{1.2}\Delta^2 u_x + \frac{(n+1)n(n-1)}{1.2.3}\Delta^3 u_x + \ldots$$

the form of which shews that the theorem remains true for the next greater value of n, therefore for the value of n still succeeding, and so on *ad infinitum*. But it is true for $n=1$, and therefore for all positive integer values of n whatever.

8. We proceed to demonstrate the same theorem by the method of generating functions.

Definition. If $\phi(t)$ is capable of being developed in a series of powers of t, the general term of the expansion being represented by $u_x t^x$, then $\phi(t)$ is said to be the generating function of u_x. And this relation is expressed in the form

$$\phi(t) = Gu_x.$$

Thus we have

$$e^t = G\frac{1}{1.2\ldots x},$$

since $\dfrac{1}{1.2\ldots x}$ is the coefficient of t^x in the development of e^t.

In like manner

$$\frac{e^t}{t} = G\frac{1}{1.2\ldots(x+1)},$$

since $\dfrac{1}{1.2\ldots(x+1)}$ is the coefficient of t^x in the development of the first member.

And generally, if $Gu_x = \phi(t)$, then

$$Gu_{x+1} = \frac{\phi(t)}{t} \;\ldots\ldots\; Gu_{x+n} = \frac{\phi(t)}{t^n} \qquad (2).$$

Hence therefore

$$Gu_{x+1} - Gu_x = \left(\frac{1}{t} - 1\right)\phi(t).$$

But the first member is obviously equal to $G\Delta u_x$, therefore

$$G\Delta u_x = \left(\frac{1}{t} - 1\right)\phi(t) \qquad (3).$$

And generally

$$G\Delta^n u_x = \left(\frac{1}{t} - 1\right)^n \phi(t) \qquad (4).$$

To apply these theorems to the problem under consideration we have, supposing still $Gu_x = \phi(t)$,

$$\begin{aligned}
Gu_{x+n} &= \left(\frac{1}{t}\right)^n \phi(t) \\
&= \left\{1 + \left(\frac{1}{t} - 1\right)\right\}^n \phi(t) \\
&= \phi(t) + n\left(\frac{1}{t} - 1\right)\phi(t) + \frac{n(n-1)}{1.2}\left(\frac{1}{t} - 1\right)^2 \phi(t) + \dots \\
&= Gu_x + nG\Delta u_x + \frac{n(n-1)}{2} G\Delta^2 u_x + \dots \\
&= G\left\{u_x + n\Delta u_x + \frac{n(n-1)}{2}\Delta^2 u_x + \dots\right\}.
\end{aligned}$$

Hence

$$u_{x+n} = u_x + n\Delta u_x + \frac{n(n-1)}{2}\Delta^2 u_x + \dots$$

which agrees with (1).

Although on account of the extensive use which has been made of the method of generating functions, especially by the older analysts, we have thought it right to illustrate its general principles, it is proper to notice that there exists an objection in point of scientific order to the employment of the method for the demonstration of the direct theorems of the Calculus of Finite Differences; viz. that G is, from its very nature, a symbol of inversion (*Diff. Equations*, p. 375, 1st Ed.). In applying it, we do not perform a direct and definite operation, but seek the answer to a question, viz. What is that function which, on performing the direct operation of development, produces terms possessing coefficients of a certain form? and this is a question which admits of an infinite variety of answers according to the extent of the development and the kind of indices supposed admissible. Hence the distributive property of the symbol G, as virtually employed in the above example, supposes limitations which are not implied in the mere definition of the symbol. It must be supposed to have reference to the same system of indices in the one member as in the other; and though, such conventions being supplied, it becomes a strict method of proof, its indirect character still remains*.

9. We proceed to the last of the methods referred to in Art. 6, viz. that which is founded upon the study of the ultimate laws of the operations involved. In addition to the symbol Δ, we shall introduce a symbol E to denote the operation of giving to x in a proposed subject function the increment unity;—its definition being

$$Eu_x \equiv u_{x+1}. \qquad (1).$$

Laws and Relations of the symbols E, Δ and $\frac{d}{dx}$.

1st. The symbol Δ is *distributive* in its operation. Thus

$$\Delta(u_x + v_x + \ldots) = \Delta u_x + \Delta v_x + \ldots \qquad (2).$$

For

$$\begin{aligned}\Delta(u_x + v_x + \ldots) &= u_{x+1} + v_{x+1} \ldots - (u_x + v_x \ldots)\\ &= u_{x+1} - u_x + v_{x+1} - v_x \ldots\\ &= \Delta u_x + \Delta v_x \ldots\end{aligned}$$

In like manner we have

$$\Delta(u_x - v_x + \ldots) = \Delta u_x - \Delta v_x + \ldots \qquad (3).$$

2ndly. The symbol Δ is *commutative* with respect to any constant coefficients in the terms of the subject to which it is applied. Thus a being constant,

$$\begin{aligned}\Delta a u_x &= a u_{x+1} - a u_x\\ &= a\Delta u_x \qquad (4).\end{aligned}$$

And from this law in combination with the preceding one, we have, $a, b, \ldots$ being constants,

$$\Delta(a u_x + b v_x + \ldots) = a\Delta u_x + b\Delta v_x + \ldots \qquad (5).$$

* The student can find instances of the use of Generating Functions in Lacroix, *Diff. and Int. Cal.* III. 322. Examples of a fourth method, at once elegant and powerful, due originally to Abel, are given in Grunert's *Archiv.* XVIII. 381.

3rdly. The symbol Δ obeys the index law expressed by the equation

$$\Delta^m \Delta^n u_x = \Delta^{m+n} u_x \qquad (6),$$

m and n being positive indices. For, by the implied definition of the index m,

$$\begin{aligned} \Delta^m \Delta^n u_x &= (\Delta\Delta \ldots m \text{ times})\,(\Delta\Delta \ldots n \text{ times})\, u_x \\ &= \{\Delta\Delta \ldots (m+n) \text{ times}\}\, u_x \\ &= \Delta^{m+n} u_x. \end{aligned}$$

These are the primary laws of combination of the symbol Δ. It will be seen from these that Δ combines with Δ and with constant quantities, as symbols of quantity combine with each other. Thus, $(\Delta + a)\, u$ denoting $\Delta u + au$, we should have, in virtue of the first two of the above laws,

$$\begin{aligned} (\Delta + a)(\Delta + b)\, u &= \{\Delta^2 + (a+b)\,\Delta + ab\}\, u \\ &= \Delta^2 u + (a+b)\,\Delta u + abu \qquad (7), \end{aligned}$$

the developed result of the combination $(\Delta + a)(\Delta + b)$ being in form the same as if Δ were a symbol of quantity.

The index law (6) is virtually an expression of the formal consequences of the truth that Δ denotes an operation which, performed upon any function of x, converts it into another function of x upon which *the same operation may be repeated.* Perhaps it might with propriety be termed the law of repetition;—as such it is common to all symbols of operation, except such, if such there be, as so alter the nature of the subject to which they are applied, as to be incapable of repetition*. It was however necessary that it should be distinctly noticed, because it constitutes a part of the formal ground of the general theorems of the calculus.

The laws which have been established for the symbol Δ are even more obviously true for the symbol E. The two symbols are connected by the equation

$$E = 1 + \Delta,$$

* For instance, if ϕ denote an operation which, when performed on *two* quantities x, y, gives a single function X, it is an operation incapable of repetition in the sense of the text, since $\phi^2(x, y) \equiv \phi(X)$ is unmeaning. But if it be taken to represent an operation which when performed on x, y, gives the *two* functions X, Y, it is capable of repetition since $\phi^2(x, y) \equiv \phi(X, Y)$, which has a definite meaning. In this case it obeys the index law.

since

$$Eu_x = u_x + \Delta u_x = (1 + \Delta)\, u_x \qquad (8),$$

and they are connected with $\dfrac{d}{dx}$ by the relation

$$E = \epsilon^{\frac{d}{dx}} \qquad (9),$$

founded on the symbolical form of Taylor's theorem. For

$$Eu_x = u_{x+1} = u_x + \frac{du_x}{dx} + \frac{1}{2}\frac{d^2u_x}{dx^2} + \frac{1}{2\,.\,3}\frac{d^3u_x}{dx^3} + \ldots$$

$$= \left(1 + \frac{d}{dx} + \frac{1}{2}\frac{d^2}{dx^2} + \frac{1}{2\,.\,3}\frac{d^3}{dx^3} + \ldots\right) u_x$$

$$= \epsilon^{\frac{d}{dx}}\, u_x.$$

It thus appears that E, Δ, and $\dfrac{d}{dx}$, are connected by the two equations

$$E = 1 + \Delta = \epsilon^{\frac{d}{dx}} \qquad (10),$$

and from the fact that E and Δ are thus both expressible by means of $\dfrac{d}{dx}$ we might have inferred that the symbols E, Δ, and $\dfrac{d}{dx}$ * combine each with itself, with constant quantities, and with each other, as if they were individually symbols of quantity. (*Differential Equations*, Chapter XVI.)

10. In the following section these principles will be applied to the demonstration of what may be termed the direct general theorems of the Calculus of Differences. The conditions of their inversion, i.e. of their extension to cases in which symbols of operation occur under negative indices, will

* In place of $\dfrac{d}{dx}$ we shall often use the symbol D. The equations will then be $E = 1 + \Delta = \epsilon^D$, a form which has the advantage of not assuming that the independent variable has been denoted by x.

be considered, so far as may be necessary, in subsequent chapters.

Ex. 1. To develope u_{x+n} in a series consisting of u_x and its successive differences (Ex. of Art. 7, resumed).

By definition

$$u_{x+1} = Eu_x, \quad u_{x+2} = E^2 u_x, \ \ldots.$$

Therefore

$$u_{x+n} = E^n u_x = (1+\Delta)^n u_x \qquad (1),$$

$$= \left\{1 + n\Delta + \frac{n(n-1)}{2}\Delta^2 + \frac{n(n-1)(n-2)}{2.3}\Delta^3 \ldots\right\} u_x$$

$$= u_x + n\Delta u_x + \frac{n(n-1)}{2}\Delta^2 u_x + \frac{n(n-1)(n-2)}{2.3}\Delta^3 u_x + \ldots \quad (2).$$

Ex. 2. To express $\Delta^n u_x$ in terms of u_x and its successive values.

Since $\Delta u_x = u_{x+1} - u_x = Eu_x - u_x$, we have

$$\Delta u_x = (E-1)u_x,$$

and as, the operations being performed, each side remains a function of x,

$$\Delta^n u_x = (E-1)^n u_x$$

$$= \left\{E^n - nE^{n-1} + \frac{n(n-1)}{1.2}E^{n-2} - \ldots\right\} u_x.$$

Hence, interpreting the successive terms,

$$\Delta^n u_x = u_{x+n} - nu_{x+n-1} + \frac{n(n-1)}{1.2} u_{x+n-2} \ldots + (-1)^n u_x \quad (3).$$

Of particular applications of this theorem those are the most important which result from supposing $u_x \equiv x^m$.

We have

$$\Delta^n x^m = (x+n)^m - n(x+n-1)^m + \frac{n(n-1)}{1.2}(x+n-2)^m - \ldots \quad (4).$$

Now let the notation $\Delta^n 0^m$ be adopted to express what the first member of the above equation becomes when $x = 0$; then

$$\Delta^n 0^m = n^m - n(n-1)^m + \frac{n(n-1)(n-2)^m}{1.2} - \frac{n(n-1)(n-2)(n-3)^m}{1.2.3} + \dots \qquad (5).$$

The systems of numbers expressed by $\Delta^n 0^m$ are of frequent occurrence in the theory of series*.

From (2) Art. 1, we have

$$\Delta^n 0^n = 1.2 \dots n,$$

and, equating this with the corresponding value given by (5), we have

$$1.2 \dots n = n^n - n(n-1)^n + \frac{n(n-1)}{1.2}(n-2)^n - \dots \qquad (6)\dagger.$$

Ex. 3. To obtain developed expressions for the nth difference of the product of two functions u_x and v_x.

Since

$$\Delta u_x v_x = u_{x+1} . v_{x+1} - u_x v_x$$
$$= Eu_x . E' v_x - u_x v_x,$$

where E applies to u_x alone, and E' to v_x alone, we have

$$\Delta u_x v_x = (EE' - 1)\, u_x v_x,$$

and generally

$$\Delta^n u_x v_x = (EE' - 1)^n\, u_x v_x \qquad (7).$$

It now only remains to transform, if needful, and to develope the operative function in the second member according to the nature of the expansion required.

Thus if it be required to express $\Delta^n u_x v_x$ in ascending differ-

* A very simple method of calculating their values will be given in Ex. 8 of this chapter.

† This formula is of use in demonstrating Wilson's Theorem, that $1 + \underline{|n-1}$ is divisible by n when n is a prime number.

ences of v_x, we must change E' into $\Delta' + 1$, regarding Δ' as operating only on v_x. We then have

$$\begin{aligned}\Delta^n u_x v_x &= \{E(1+\Delta') - 1\}^n u_x v_x \\ &= (\Delta + E\Delta')^n u_x v_x \\ &= \left\{\Delta^n + n\Delta^{n-1}E\Delta' + \frac{n(n-1)}{1.2}\Delta^{n-2}E^2\Delta'^2 + \dots\right\} u_x v_x.\end{aligned}$$

Remembering then that Δ and E operate only on u_x and Δ' only on v_x, and that the accent on the latter symbol may be dropped when that symbol only precedes v_x, we have

$$\Delta^n u_x v_x = \Delta^n u_x \,.\, v_x + n\Delta^{n-1} u_{x+1} \,.\, \Delta v_x + \frac{n(n-1)}{1.2}\Delta^{n-2} u_{x+2} \,.\, \Delta^2 v_x + \dots \quad (8),$$

the expansion required.

As a particular illustration, suppose $u_x \equiv a^x$. Then, since

$$\begin{aligned}\Delta^{n-r} u_{x+r} &= \Delta^{n-r} a^{x+r} = a^r \Delta^{n-r} a^x \\ &= a^{x+r}(a-1)^{n-r}, \text{ by (14), Art. 2,}\end{aligned}$$

we have

$$\Delta^n a^x v_x = a^x \left\{(a-1)^n v_x + n(a-1)^{n-1} a\Delta v_x + \frac{n(n-1)}{2}(a-1)^{n-2} a^2 \Delta^2 v_x + \dots\right\} \quad (9).$$

Again, if the expansion is to be ordered according to successive values of v_x, it is necessary to expand the untransformed operative function in the second member of (7) in ascending powers of E' and develope the result. We find

$$\Delta^n u_x v_x = (-1)^n \left\{u_x v_x - n u_{x+1} v_{x+1} + \frac{n(n-1)}{2} u_{x+2} v_{x+2} - \dots\right\} \quad (10).$$

Lastly, if the expansion is to involve only the differences of u_x and v_x, then, changing E into $1 + \Delta$, and E' into $1 + \Delta'$, we have

$$\Delta^n u_x v_x = (\Delta + \Delta' + \Delta\Delta')^n u_x v_x \quad (11),$$

and the symbolic trinomial in the second member is now to be developed and the result interpreted.

Ex. 4. To express $\Delta^n u_x$ in terms of the differential coefficients of u_x.

By (10), Art. 9, $\Delta = \epsilon^{\frac{d}{dx}} - 1$. Hence

$$\Delta^n u_x = (\epsilon^{\frac{dt}{dx}} - 1)^n u_x \qquad (12).$$

Now t being a symbol of quantity, we have

$$(\epsilon^t - 1)^n = \left(t + \frac{t^2}{1.2} + \frac{t^2}{1.2.3} + \dots\right)^n \qquad (13),$$

$$= t^n + A_1 t^{n+1} + A_2 t^{n+2} + \dots,$$

on expansion, A_1, A_2, being numerical coefficients. Hence

$$(\epsilon^{\frac{d}{dx}} - 1)^n = \left(\frac{d}{dx}\right)^n + A_1\left(\frac{d}{dx}\right)^{n+1} + A_2\left(\frac{d}{dx}\right)^{n+2} + \dots,$$

and therefore

$$\Delta_n u_x = \left(\frac{d}{dx}\right)^n u_x + A_1\left(\frac{d}{dx}\right)^{n+1} u_x + A_2\left(\frac{d}{dx}\right)^{n+2} u_x + \dots \qquad (14).$$

The coefficients A_1, A_2,... may be determined in various ways, the simplest in principle being perhaps to develope the right-hand member of (13) by the polynomial theorem, and then seek the aggregate coefficients of the successive powers of t. But the expansion may also be effected with complete determination of the constants by a remarkable secondary form of Maclaurin's theorem, which we shall proceed to demonstrate.

Secondary form of Maclaurin's Theorem.

PROP. *The development of $\phi(t)$ in positive and integral powers of t, when such development is possible, may be expressed in the form*

$$\phi(t) = \phi(0) + \phi\left(\frac{d}{d0}\right)0 \, . \, t + \phi\left(\frac{d}{d0}\right)0^2 \, . \, \frac{t^2}{1.2}$$

$$+ \phi\left(\frac{d}{d0}\right)0^3 \, . \, \frac{t^3}{1.2.3} + \dots$$

where $\phi\left(\frac{d}{d0}\right)0^m$ denotes what $\phi\left(\frac{d}{dx}\right)x^m$ becomes when $x = 0$.

First, we shall shew that if $\phi(x)$ and $\psi(x)$ are any two functions of x admitting of development in the form $a+bx+cx^2+\ldots$,

then $$\phi\left(\frac{d}{dx}\right)\psi(x)=\psi\left(\frac{d}{dx}\right)\phi(x) \qquad (15),$$

provided that x be made equal to 0, after the implied operations are performed.

For, developing all the functions, each member of the above equation is resolved into a series of terms of the form $A\left(\frac{d}{dx}\right)^m x^n$, while in corresponding terms of the two members the order of the indices m and n will be reversed.

Now $\left(\frac{d}{dx}\right)^m x^n$ is equal to 0 if m is greater than n, to $1.2\ldots n$ if m is equal to n, and again to 0 if m is less than n and at the same time x equal to 0; for in this case x^{n-m} is a factor. Hence if $x=0$,

$$\left(\frac{d}{dx}\right)^m x^n=\left(\frac{d}{dx}\right)^n x^m,$$

and therefore under the same condition the equation (15) is true, or, adopting the notation above explained,

$$\phi\left(\frac{d}{d0}\right)\psi(0)=\psi\left(\frac{d}{d0}\right)\phi(0) \qquad (16).$$

Now by Maclaurin's theorem in its known form

$$\phi(t)=\phi(0)+\frac{d}{d0}\phi(0).t+\frac{d^2}{d0^2}\phi(0).\frac{t^2}{1.2}+\ldots \qquad (17).$$

Hence, applying the above theorem of reciprocity,

$$\phi(t)=\phi(0)+\phi\left(\frac{d}{d0}\right)0.t+\phi\left(\frac{d}{d0}\right)0^2.\frac{t^2}{1.2}+\ldots \qquad (18),$$

the secondary form in question. The two forms of Maclaurin's theorem (17), (18) may with propriety be termed *conjugate.*

A simpler proof of the above theorem (which may be more shortly written $\phi(t) = \phi(D)\,\epsilon^{0.t}$) is obtained by regarding it as a particular case of Herschel's theorem, viz.

$$\phi(\epsilon^t) = \phi(1) + \phi(E)\,0\,.\,t + \phi(E)\,0^2\,.\,\frac{t^2}{1\,.\,2} + \dots \qquad (19),$$

or, symbolically written, $\phi(\epsilon^t) = \phi(E)\,\epsilon^{0.t}$.* The truth of the last theorem is at once rendered evident by assuming $A_n\epsilon^{nt}$ to be any term in the expansion of $\phi(\epsilon^t)$ in powers of ϵ^t. Then since $A_n\,\epsilon^{nt} \equiv A_n E^n\,\epsilon^{0.t}$ the identity of the two series is evident.

But $$\phi(t) \equiv \phi(\log \epsilon^t) = \phi(\log E)\,\epsilon^{0.t}$$

(by Herschel's theorem)

$$\equiv \phi(D)\,\epsilon^{0.t},$$

which is the secondary form of Maclaurin's theorem.

As a particular illustration suppose $\phi(t) \equiv (\epsilon^t - 1)^n$, then by means of either of the above theorems we easily deduce

$$(\epsilon^t - 1)^n = \Delta^n 0\,.\,t + \Delta^n 0^2\,.\,\frac{t^2}{1\,.\,2} + \Delta^n 0^3\,.\,\frac{t^3}{1\,.\,2\,.\,3} + \dots.$$

But $\Delta^n 0^m$ is equal to 0 if m is less than n and to $1\,.\,2\,.\,3 \dots n$ if m is equal to n, (Art. 1). Hence

$$(\epsilon^t - 1)^n = t^n + \frac{\Delta^n 0^{n+1}}{1\,.\,2 \dots (n+1)}\,.\,t^{n+1} + \frac{\Delta^n 0^{n+1}}{1\,.\,2 \dots (n+2)}\,.\,t^{n+2} + \dots \qquad (20).$$

Hence therefore since $\Delta^n u = (\epsilon^{\frac{d}{dx}} - 1)^n u$ we have

$$\Delta^n u = \frac{d^n u}{dx^n} + \frac{\Delta^n 0^{n+1}}{1.2 \dots (n+1)}\,.\,\frac{d^{n+1} u}{dx^{n+1}} + \frac{\Delta^n 0^{n+2}}{1.2 \dots (n+2)}\,\frac{d^{n+2} u}{dx^{n+2}} + \dots \qquad (21),$$

the theorem sought.

The reasoning employed in the above investigation proceeds upon the assumption that n is a positive integer. The

* Since both Δ and D performed on a constant produce as result zero, it is obvious that $\phi(D)\,C = \phi(0)\,C = \phi(\Delta)\,C$, and $\phi(E)\,C = \phi(1)\,C$. It is of course assumed throughout that the coefficients in ϕ are constants.

very important case in which $n=-1$ will be considered in another chapter of this work.

Ex. 5. *To express* $\dfrac{d^n u}{dx^n}$ *in terms of the successive differences of* u.

Since $\epsilon^{\frac{d}{dx}}=1+\Delta$, we have

$$\frac{d}{dx}=\log(1+\Delta),$$

therefore
$$\left(\frac{d}{dx}\right)^n=\left\{\log(1+\Delta)\right\}^n \qquad (22),$$

and the right-hand member must now be developed in ascending powers of Δ.

In the particular case of $n=1$, we have

$$\frac{du}{dx}=\Delta u-\frac{\Delta^2 u}{2}+\frac{\Delta^3 u}{3}-\frac{\Delta^4 u}{4}+\ \ldots \qquad (23).$$

11. It would be easy, but it is needless, to multiply these general theorems, some of those above given being valuable rather as an illustration of principles than for their intrinsic importance. We shall, however, subjoin two general theorems, of which (21) and (23) are particular cases, as they serve to shew how striking is the analogy between the parts played by factorials in the Calculus of Differences and powers in the Differential Calculus.

By Differential Calculus we have

$$u_{x+t}=u_x+t\,.\frac{du_x}{dx}+\frac{t^2}{1\,.\,2}\cdot\frac{d^2u_x}{dx^2}+\ldots.$$

Perform $\phi(\Delta)$ on both sides (Δ having reference to t alone), and subsequently put $t=0$. This gives

$$\phi(\Delta)\,u_x=u_x\,.\,\phi(0)+\phi(\Delta)\,0\,.\frac{du_x}{dx}+\frac{\phi(\Delta)\,0^2}{1\,.\,2}\cdot\frac{d^2u_x}{dx^2}+\ \ldots \qquad (24),$$

of which (21) is a particular case.

By (2) we have

$$u_{x+t} = u_x + t \,.\, \Delta u_x + \frac{t^{(2)}}{1\,.\,2} \Delta^2 u_x + \ldots.$$

Perform $\phi\left(\frac{d}{dt}\right)$ on each side, and subsequently put $t = 0$;

$$\therefore \phi\left(\frac{d}{dx}\right) u_x = u_x \,.\, \phi(0) + \phi\left(\frac{d}{d0}\right) 0 \,.\, \Delta u_x$$

$$+ \phi\left(\frac{d}{d0}\right) 0^{(2)} \,.\, \frac{\Delta^2 u_x}{1\,.\,2} + \ldots \qquad (25),$$

of which (23) is a particular case.

12. We have seen in Art. 9 that the symbols Δ, E and $\frac{d}{dx}$ or D have, with certain restrictions, the same laws of combination as constants. It is easy to see that, in general, these laws will hold good when they combine with other symbols of operation provided that these latter also obey the above-mentioned laws. By these means the Calculus of Finite Differences may be made to render considerable assistance to the Infinitesimal Calculus, especially in the evaluation of Definite Integrals. We subjoin two examples of this; further applications of this method may be seen in a *Mémoire* by *Cauchy* (*Journal Polytechnique*, Vol. XVII.).

Ex. 6. To shew that $B(m+1, n) = (-1)^m \Delta^m \frac{1}{n}$, where m is a positive integer.

We have $$\frac{1}{n} = \int_0^\infty \epsilon^{-nx} dx;$$

$$\therefore \Delta^m \frac{1}{n} = \Delta^m \int_0^\infty \epsilon^{-nx} dx = \int_0^\infty \Delta^m \epsilon^{-nx} dx$$

$$= \int_0^\infty \epsilon^{-nx} (\epsilon^{-x} - 1)^m \, dx$$

$$= \int_0^1 z^{n-1} (z-1)^m \, dz \text{ (assuming } z = \epsilon^{-x})$$

$$= (-1)^m B(m+1, n).$$

Ex. 7. Evaluate $u=\int_0^\infty \Delta^m \frac{z^{a+1}}{z^2+n^2}dz$, m being a positive integer greater than a; Δ relating to n alone.

Let 2κ be the even integer next greater than $a+1$, then

$$\frac{z^{a+1}}{z^2+n^2} \equiv z^{a-2\kappa+1}\left\{\frac{z^{2\kappa}-n^{2\kappa}}{z^2+n^2}+\frac{n^{2\kappa}}{z^2+n^2}\right\} \qquad (26).$$

Now the first member of the right-hand side of (26) is a rational integral function of n of an order lower than m. It therefore vanishes when the operation Δ^m is performed on it. We have therefore

$$u=\int_0^\infty \Delta^m \frac{n^{2\kappa}}{z^2+n^2} z^{a-2\kappa+1}dz = \Delta^m \int_0^\infty \frac{n^{2\kappa}}{z^2+n^2} z^{a-2\kappa+1}dz \qquad (27)$$

$$=\frac{1}{2}\Delta^m n^a \int_0^\infty \frac{y^{\frac{a}{2}-\kappa}}{y+1}dy \text{ (assuming } z^2=n^2y)$$

$$=\frac{1}{2}\frac{\pi}{\sin\left(\frac{a}{2}-\kappa+1\right)\pi}.\Delta^m n^a$$

(Tod. *Int. Cal.* Art. 255, 3rd Ed.)

$$=-\frac{1}{2}(-1)^\kappa \frac{\pi}{\sin\frac{a\pi}{2}}\Delta^m n^a.$$

This example illustrates strikingly the nature and limits of the commutability of order of the operations $\int$ and Δ. Had we changed the order (as in (27)) without previously preparing the quantity under the sign of integration, we should have had

$$\Delta^m n^a . \int_0^\infty \frac{y^{\frac{a}{2}}}{y+1}dy,$$

which is infinite if a be positive.

The explanation of this singularity is as follows:—

If we write for Δ^m its equivalent $(E-1)^m$ and expand the latter, we see that $\int_0^\infty \Delta^m \phi(x,n)\,dx$ expresses the integral of a quantity of $m+1$ terms of the form $A_p\,\phi(x, n+p)$, while $\Delta^m\int_0^\infty \phi(x, n)\,dx$ expresses the sum of $m+1$ separate integrals, each having under the integral sign one of the terms of the above quantity. Where each term separately integrated gives a finite result, it is of course indifferent which form is used, but where, as in the case before us, two or more would give infinity as result the second form cannot be used.

13. Ex. 8. To shew that

$$\phi(E)\,0^n = E\phi'(E)\,0^{n-1}. \qquad (28).$$

Let $A_r E^r\,0^n$ and $ErA_r E^{r-1}\,0^{n-1}$ be corresponding terms of the two expansions in (28). Then, since each of them equals $A_r r^n$, the identity of the two series is manifest.

Since $E \equiv 1 + \Delta$ the theorem may also be written

$$\phi(\Delta)\,0^n = E\phi'(\Delta)\,0^{n-1},$$

and under this form it affords the simplest mode of calculating the successive values of $\Delta^m 0^n$. Putting $\phi(\Delta) \equiv \Delta^m$, we have

$$\Delta^m 0^n = E\,.\,m\Delta^{m-1}0^{n-1} = m\,(\Delta^{m-1}0^{n-1} + \Delta^m 0^{n-1}),$$

and the differences of 0^n can be at once calculated from those of 0^{n-1}.

Other theorems about the properties of the remarkable set of numbers of the form $\Delta^m 0^n$ will be found in the accompanying exercises. Those desirous of further information on the subject may consult the papers of Mr J. Blissard and M. Worontzof in the *Quarterly Journal of Mathematics*, Vols. VIII. and IX.

EXERCISES.

1. Find the first differences of the following functions:

$$2^x \sin\frac{a}{2^x}, \qquad \tan\frac{a}{2^x}, \qquad \cot(2^x a).$$

2. Shew that

$$\Delta \frac{u_x}{v_x} = \frac{v_x \Delta u_x - u_x \Delta v_x}{v_x v_{x+1}}.$$

3. Prove the following theorems:

$$\Delta^n 0^{n+1} = \frac{n(n+1)}{2} \Delta^n 0^n$$

$$\Delta 0^x - \frac{\Delta^2}{2} 0^x + \&c. = 0 \; (x > 1)$$

$$\Delta^m 0^n + n\Delta^m 0^{n-1} + \frac{n(n-1)}{1.2} \Delta^m 0^{n-2} + \ldots + \frac{\lfloor n}{\lfloor n-m} = \frac{\Delta^{m+1} 0^{n+1}}{m+1}.$$

$$\phi(E^n) 0^x = n^x \phi(E) 0^x.$$

4. Shew that, if m be less than r,

$$\{1 + \log E\}^r 0^m = r(r-1) \ldots\ldots (r-m+1).$$

5. Express the differential coefficient of a factorial in factorials. Ex. $x^{(m)}$.

6. Shew that

$$\Delta^n 0^n, \; \Delta^n 0^{n+1} \ldots\ldots$$

form a recurring series, and find its scale of relation.

7. If $P_x^n \equiv \dfrac{\Delta^n 0^x}{\lfloor n}$ shew that

$$P_x^n = P_{x-1}^{n-1} + nP_{x-1}^n.$$

8. Shew that

$$u_0 + u_1 x + \frac{u_2 x^2}{1.2} + \&c. = \epsilon^x \left\{ u_0 + x\Delta u_0 + \frac{x^2 \Delta^2 u_0}{1.2} + \ldots \right\}$$

What class of series would the above theorem enable us to convert from a slow to a rapid convergence?

9. Shew that

$$\epsilon^{\epsilon^t} = \epsilon\left\{1 + (\epsilon^{\Delta}0)\, t + (\epsilon^{\Delta}0^2)\frac{t^2}{1\,.\,2} + \cdots\right\},$$

and hence calculate the first four terms of the expression.

10. If $P_n = \frac{1}{1\,.\,n} + \frac{1}{2\,(n-1)} + \ldots + \frac{1}{n\,.\,1}$, shew that

$$(P_1\Delta^2 - P_2\Delta^3 + \ldots)\, 0^m = 0 \text{ if } m > 2.$$

Prove that

$$\{\log E\}^n\, 0^m = 0,$$

unless $m = n$ when it is equal to $\lfloor n$.

11. Prove that

$$\frac{1}{x^2 + nx} = x^{(-2)} + (1-n)\, x^{-3} + (1-n)(2-n)\, x^{-4} + \ldots.$$

12. If $x = \epsilon^\theta$, prove that

$$\left(\frac{d}{d\theta}\right)^n = \frac{\Delta 0^n}{1}\,.\,x\frac{d}{dx} + \frac{\Delta^2 0^n}{1\,.\,2}\, x^2\frac{d^2}{dx^2} + \frac{\Delta^3 0^n}{1\,.\,2\,.\,3}\, x^3\frac{d^3}{dx^3} + \ldots$$

13. If $\Delta u_{x,\,y} \equiv u_{x+1,\,y+1} - u_{x,\,y}$ and if $\Delta^n u_{x,\,y}$ be expanded in a series of differential coefficients of $u_{x,\,y}$, shew that the general term will be

$$\frac{\Delta^n 0^{p+q}}{\Delta^p 0^p \times \Delta^q 0^q}\,.\,\frac{d^{p+q}\, u_{x,\,y}}{dx^p dy^q}.$$

14. Express $\Delta^n x^m$ in a series of terms proceeding by powers of x by means of the differences of the powers of 0.

By means of the same differences, find a finite expression for the infinite series

$$1^m\,.\,x - 3^m\,.\,\frac{x^3}{\lfloor 3} + 5^m\,.\,\frac{x^5}{\lfloor 5} + \ldots$$

where m is a positive integer, and reduce the result when

$$m = 4.$$

15. Prove that

$$F(E)\, a^x \phi x = a^x F(aE)\, \phi x,$$

$$(x\Delta)^{(n)} u_x = (x+n-1)^{(n)} \Delta u_x,$$

$$f(x\Delta)\,(xE)^m u_x = (xE)^m f(x\Delta + m)\, u_x,$$

and find the analogous theorems in the Infinitesimal Calculus.

16. Find u_x from the equations

$$(1)\quad Gu_x = \frac{1-\sqrt{1-4t^2}}{2t};$$

$$(2)\quad Gu_x = f(\epsilon^t).$$

17. Find a symbolical expression for the nth difference of the product of any number of functions in terms of the differences of the separate functions, and deduce Leibnitz's theorem therefrom.

18. If P_n be the number of ways in which a polygon of n sides can be divided into triangles by its diagonals, and $t^2\phi(t) \equiv GP_n$, shew that

$$\phi(t) = \sqrt{\frac{\phi(t)-1}{t}}.$$

*19. Shew that

$$\int_0^\infty \epsilon^{-nx}(\epsilon^{-x}-1)^m x^{a-1}\, dx = \Gamma a\, \Delta^m n^{-a},$$

n and a being positive quantities.

*20. Shew that

$$\int_0^\infty \frac{\sin 2nx \sin^m x}{x^{a+1}}\, dx = \frac{\pi \Delta^m (2n-m)^a}{2^{m+1}\,\Gamma(a+1)\cos\dfrac{a+m}{2}\pi},$$

if $2n > m > a$ all being positive.

* In Questions 19 and 20 Δ acts on n alone.

Hence, shew that $\int_0^\infty \sin 2nx \, . \, \frac{\sin^m x}{x^{m+1}} \, . \, dx$ is constant for all values of n between $\frac{m}{2}$ and ∞.

21. Shew that if p be a positive integer

$$\int_0^\infty \epsilon^{-\kappa x} \, . \, \sin^{2p} x \, . \, dx = \frac{1 \, . \, 2 \, . \, 3 \ldots\ldots 2p}{\kappa \, (\kappa^2 + 4) \ldots\ldots (\kappa^2 + 4p^2)} \, .$$

(Bertrand, *Cal. Int.* p. 185.)

22. Shew that

$$\Delta^{n-1} \, 2^p = \frac{\Delta^n \, 1^{p+1} + \Delta^{n-1} \, 1^p}{n+1} \, .$$

23. Demonstrate the formula

$$\Delta^n \, 1^{p+1} = (n+1) \, \Delta^n \, 1^p + n \Delta^{n-1} \, 1^p,$$

and apply it to construct a table of the differences of the powers of unity up to the fifth power.

CHAPTER III.

ON INTERPOLATION, AND MECHANICAL QUADRATURE.

1. THE word interpolate has been adopted in analysis to denote primarily the interposing of missing terms in a series of quantities supposed subject to a determinate law of magnitude, but secondarily and more generally to denote the calculating, under some hypothesis of law or continuity, of *any* term of a series from the values of any other terms supposed given.

As no series of *particular* values can determine a law, the problem of interpolation is an indeterminate one. To find an analytical expression of a function from a limited number of its numerical values corresponding to given values of its independent variable x is, in Analysis, what in Geometry it would be to draw a continuous curve through a number of given points. And as in the latter case the number of possible curves, so in the former the number of analytical expressions satisfying the given conditions, is infinite. Thus the form of the function—the species of the curve—must be assumed *a priori*. It may be that the evident character of succession in the values observed indicates what kind of assumption is best. If for instance these values are of a periodical character, circular functions ought to be employed. But where no such indications exist it is customary to assume for the general expression of the values under consideration a rational and integral function of x, and to determine the coefficients by the given conditions.

This assumption rests upon the supposition (a supposition however actually verified in the case of all tabulated functions) that the successive orders of differences rapidly diminish. In the case of a rational and integral function of x of the nth degree it has been seen that differences of the $n+1^{\text{th}}$

and of all succeeding orders vanish. Hence if in any other function such differences become very small, that function may, quite irrespectively of its form, be approximately represented by a function which is rational and integral. Of course it is supposed that the value of x for which that of the function is required is not very remote from those, or from some of those, values for which the values of the function are given. The same assumption as to the form of the unknown function and the same condition of limitation as to the use of that form flow in an equally obvious manner from the expansion in Taylor's theorem.

2. The problem of interpolation assumes different forms, according as the values given are equidistant, i.e. correspondent to equidifferent values of the independent variable, or not. But the solution of all its cases rests upon the same principle. The most *obvious* mode in which that principle can be applied is the following. If for n values $a, b, \dots$ of an independent variable x the corresponding value $u_a, u_b, \dots$ of an unknown function of x represented by u_x, are given, then, assuming as the approximate general expression of u_x,

$$u_x = A + Bx + Cx^2 \dots + Ex^{n-1} \qquad (1),$$

a form which is rational and integral and involves n arbitrary coefficients, the data in succession give

$$u_a = A + Ba + Ca^2 \dots + Ea^{n-1},$$
$$u_b = A + Bb + Cb^2 \dots + Eb^{n-1},$$
$$\dots\dots\dots\dots\dots\dots$$

a system of n linear equations which determine $A, B \dots E$. To avoid the solving of these equations other but equivalent modes of procedure are employed, all such being in effect reducible to the two following, viz. either to an application of that property of the rational and integral function in the second member of (1) which is expressed by the equation $\Delta^n u_x = 0$, or to the substitution of a different but equivalent form for the rational and integral function. These methods will be respectively illustrated in Prop. 1 and its deductions, and in Prop. 2, of the following sections.

PROP. 1. Given n consecutive equidistant values $u_0, u_1, \dots u_{n-1}$ of a function u_x, to find its approximate general expression.

By Chap. II. Art. 10,

$$u_{x+m} = u_x + m\Delta u_x + \frac{m(m-1)}{1.2}\Delta^2 u_x + \dots.$$

Hence, substituting 0 for x, and x for m, we have

$$u_x = u_0 + x\Delta u_0 + \frac{x(x-1)}{1.2}\Delta^2 u_0 + \dots.$$

But on the assumption that the proposed expression is rational and integral and of the degree $n-1$, we have $\Delta^n u_x = 0$, and therefore $\Delta^n u_0 = 0$. Hence

$$u_x = u_0 + x\Delta u_0 + \frac{x(x-1)}{1.2}\Delta^2 u_0 \dots$$
$$+ \frac{x(x-1)\dots(x-n+2)}{1.2\dots(n-1)}\Delta^{n-1} u_0 \qquad (2),$$

the expression required. It will be observed that the second member is really a rational and integral function of x of the degree $n-1$, while the coefficients are made determinate by the data.

In applying this theorem the value of x may be conceived to express the *distance* of the term sought from the first term in the series, the common distance of the terms given being taken as unity.

Ex. Given $\log 3{\cdot}14 = {\cdot}4969296$, $\log 3{\cdot}15 = {\cdot}4983106$, $\log 3{\cdot}16 = {\cdot}4996871$, $\log 3{\cdot}17 = {\cdot}5010593$; required an approximate value of $\log 3{\cdot}14159$.

Here, omitting the decimal point, we have the following table of numbers and differences:

	u_0	u_1	u_2	u_3
	4969296	4983106	4996871	5010593
Δ	13810	13765	13722	
Δ^2	−45	−43		
Δ^3	2			

The first column gives the values of u_0 and its differences up to $\Delta^3 u_0$. Now the common difference of 3·14, 3·15, ...

being taken as unity, the value of x which corresponds to 3·14159 will be ·159. Hence we have

$$u_x = 4969296 + \cdot 159 \times 13810 + \frac{(\cdot 159)(\cdot 159 - 1)}{1.2} \times (-45)$$
$$+ \frac{(\cdot 159)(\cdot 159 - 1)(\cdot 159 - 2)}{1.2.3} \times 2.$$

Effecting the calculations we find $u_x = \cdot 4971495$, which is true to the last place of decimals. Had the first difference only been employed, which is equivalent to the ordinary rule of proportional parts, there would have been an error of 3 in the last decimal.

3. When the values given and that sought constitute a series of equidistant terms, whatever may be the position of the value sought in that series, it is better to proceed as follows.

Let $u_0, u_1, u_2, \ldots u_n$ be the series. Then since, according to the principle of the method, $\Delta^n u_0 = 0$, we have by Chap. II. Art. 10,

$$u_n - nu_{n-1} + \frac{n(n-1)}{1.2} u_{n-2} - \ldots + (-1)^n u_0 = 0 \qquad (3),$$

an equation from which any one of the quantities

$$u_0, u_1, \ldots u_n$$

may be found in terms of the others.

Thus, to interpolate a term midway between two others we have

$$u_0 - 2u_1 + u_2 = 0; \ \therefore \ u_1 = \frac{u_0 + u_2}{2} \qquad (4).$$

Here the middle term is only the arithmetical mean.

To supply the middle term in a series of five, we have

$$u_0 - 4u_1 + 6u_2 - 4u_3 + u_4 = 0;$$
$$\therefore \ u_2 = \frac{4(u_1 + u_3) - (u_0 + u_4)}{6} \qquad (5).$$

Ex. Representing as is usual $\int_0^\infty \epsilon^{-\theta} \theta^{n-1} d\theta$ by $\Gamma(n)$, it is required to complete the following table by finding approximately $\log \Gamma\left(\frac{1}{2}\right)$:

n	$\log \Gamma(n)$,	n	$\log \Gamma(n)$,
$\frac{2}{12}$	·74556,	$\frac{7}{12}$	·18432,
$\frac{3}{12}$	·55938,	$\frac{8}{12}$	·13165,
$\frac{4}{12}$	·42796,	$\frac{9}{12}$	·08828,
$\frac{5}{12}$	·32788,	$\frac{10}{12}$	·05261.

Let the series of values of $\log \Gamma(n)$ be represented by $u_1, u_2, \ldots u_9$, the value sought being that of u_5. Then proceeding as before, we find

$$u_1 - 8u_2 + \frac{8.7}{1.2} u_3 - \frac{8.7.6}{1.2.3} u_4 + \ldots = 0,$$

or,

$$u_1 + u_9 - 8(u_2 + u_8) + 28(u_3 + u_7) - 56(u_4 + u_6) + 70u_5 = 0;$$

whence

$$u_5 = \frac{56(u_4 + u_6) - 28(u_3 + u_7) + 8(u_2 + u_8) - (u_1 + u_9)}{70} \qquad (6).$$

Substituting for u_1, u_2, ..., their values from the table, we find

$$\log \Gamma\left(\frac{1}{2}\right) = \cdot 24853,$$

the true value being ·24858.

To shew the gradual closing of the approximation as the number of the values given is increased, the following results are added:

Data.	Calculated value of u_5.
$u_4 \quad u_6$	.... ·25610,
$u_3, u_4 \quad u_6, u_7$	.... ·24820,
$u_2, u_3, u_4 \quad u_6, u_7, u_8,$	.... ·24865,
$u_1, u_2, u_3, u_4 \quad u_6, u_7, u_8, u_9$	.... ·24853.

4. By an extension of the same method, we may treat any case in which the terms given and sought are terms, but not consecutive terms, of a series. Thus, if u_1, u_4, u_5 were given and u_3 sought, the equations $\Delta^3 u_1 = 0$, $\Delta^3 u_2 = 0$ would give

$$u_4 - 3u_3 + 3u_2 - u_1 = 0,$$
$$u_5 - 3u_4 + 3u_3 - u_2 = 0,$$

from which, eliminating u_2, we have

$$3u_5 - 8u_4 + 6u_3 - u_1 = 0 \quad\quad (7),$$

and hence u_3 can be found. But it is better to apply at once the general method of the following Proposition.

PROP. 2. Given n values of a function which are not consecutive and equidistant, to find any other value whose place is given.

Let $u_a, u_b, u_c, \ldots u_k$ be the given values, corresponding to $a, b, c, \ldots k$ respectively as values of x, and let it be required to determine an approximate general expression for u_x.

We shall assume this expression rational and integral, Art. 1.

Now there being n conditions to be satisfied, viz. that for $x = a$, $x = b \ldots x = k$, it shall assume the respective values $u_a, u_b, \ldots u_k$, the expression must contain n constants, whose values those conditions determine.

We might therefore assume

$$u_x = A + Bx + Cx^2 \ldots + Ex^{n-1} \quad\quad (8),$$

and determine A, B, C by the linear system of equations formed by making $x = a, b \ldots k$, in succession.

The substitution of another but equivalent form for (8) enables us to dispense with the solution of the linear system.

Let

$$\begin{aligned} u_x = {} & A\,(x-b)\,(x-c)\,\ldots\,(x-k) \\ & + B\,(x-a)\,(x-c)\,\ldots\,(x-k) \\ & + C\,(x-a)\,(x-b)\,\ldots\,(x-k) \\ & + \ldots \end{aligned} \qquad (9)$$

to n terms, each of the n terms in the right-hand member wanting one of the factors $x-a$, $x-b$, ... $x-k$, and each being affected with an arbitrary constant. The assumption is legitimate, for the expression thus formed is, like that in (8), rational and integral, and it contains n undetermined coefficients.

Making $x=a$, we have

$$u_a = A\,(a-b)\,(a-c)\,\ldots\,(a-k);$$

therefore

$$A = \frac{u_a}{(a-b)\,(a-c)\,\ldots\,(a-k)}.$$

In like manner making $x=b$, we have

$$B = \frac{u_b}{(b-a)\,(b-c)\,\ldots\,(b-k)},$$

and so on. Hence, finally,

$$u_x = u_a \frac{(x-b)\,(x-c)\,\ldots\,(x-k)}{(a-b)\,(a-c)\,\ldots\,(a-k)} + u_b \frac{(x-a)\,(x-c)\,\ldots\,(x-k)}{(b-a)\,(b-c)\,\ldots\,(b-k)} \ldots$$

$$+ \;\ldots\; + u_k \frac{(x-a)\,(x-b)\,(x-c)\,\ldots}{(k-a)\,(k-b)\,(k-c)\,\ldots} \qquad (10),$$

the expression required. This is Lagrange's* theorem for interpolation.

If we assume that the values are consecutive and equidistant, i. e. that u_0, u_1 ... u_{n-1} are given, the formula becomes

$$u_x = u_{n-1} \frac{x\,(x-1)\,\ldots\,(x-n+2)}{1\,.\,2\,.\,3\,\ldots\,(n-1)} - u_{n-2} \frac{x\,(x-1)\,\ldots\,(x-n+1)}{1\,.\,1\,.\,2\,\ldots\,(n-2)}$$

$$+ \ldots$$

* *Journal de l'Ecole Polytechnique*, II. 277. The real credit of the discovery must, however, be assigned to Euler; who, in a tract entitled *De eximio usu methodi interpolationum in serierum doctrina*, had, long before this, obtained a closely analogous expression.

$$= \frac{x(x-1)\ldots(x-n+1)}{\lfloor n-1}\left\{\frac{u_{n-1}}{x-n+1} - C_1\frac{u_{n-2}}{x-n+2} + \ldots\right\} \quad (11),$$

where
$$C_r \equiv \frac{\lfloor n-1}{\lfloor r \lfloor n-1-r}.$$

This formula may be considered as conjugate to (2), and possesses the advantage of being at once written down from the observed values of u_x without our having to compute the successive differences. But this is more than compensated for in practice, especially when the number of available observations is large, by the fact that in forming the coefficients in (2) we are constantly made aware of the degree of closeness of the approximation by the smallness of the value of $\Delta^n u_0$, and can thus judge when we may with safety stop.

As the problem of interpolation, under the assumption that the function to be determined is rational and integral and of a degree not higher than the $(n-1)^{\text{th}}$, is a determinate one, the different methods of solution above exemplified lead to consistent results. All these methods are implicitly contained in that of Lagrange.

The following are particular applications of Lagrange's theorem.

5. Given any number of values of a magnitude as observed at given times; to determine approximately the values of the successive differential coefficients of that magnitude at another given time.

Let $a, b, \ldots k$ be the times of observation, $u_a, u_b, \ldots u_k$ the observed values, x the time for which the value is required, and u_x that value. Then the value of u_x is given by (10), and the differential coefficients can thence be deduced in the usual way. But it is most convenient to assume the time represented above by x as the epoch, and to regard $a, b, \ldots k$ as measured from that epoch, being negative if measured backwards. The values of $\frac{du_x}{dx}, \frac{d^2u_x}{dx^2}, \ldots$ will then be the coefficients of $x, x^2, \ldots$ in the development of the second member of (10) multiplied by $1, 1.2, 1.2.3, \ldots$ successively. Their general expressions may thus at once be found. Thus in particular we shall have

$$\frac{du_x}{dx} = \pm \frac{bc \ldots k\left(\frac{1}{b} + \frac{1}{c} \ldots + \frac{1}{k}\right)}{(a-b)(a-c)\ldots(a-k)} u_a \pm \ldots \qquad (12),$$

$$\frac{d^2u_x}{dx^2} = \mp 1 \, . \, 2 \, . \frac{bc \ldots k\left(\frac{1}{bc} + \frac{1}{bd} + \frac{1}{cd} + \ldots\right)}{(a-b)(a-c)\ldots(a-k)} u_a \mp \ldots \qquad (13).$$

Laplace's computation of the orbit of a comet is founded upon this proposition (*Mécanique Celeste*).

6. The values of a quantity, e.g. the altitude of a star at given times, are found by observation. Required at what intermediate time the quantity had another given value.

Though it is usual to consider the time as the independent variable, in the above problem it is most convenient to consider the observed magnitude as such, and the time as a function of that magnitude. Let then $a, b, c, \ldots$ be the values given by observation, $u_a, u_b, u_c, \ldots$ the corresponding times, x the value for which the time is sought, and u_x that time. Then the value of u_x is given at once by Lagrange's theorem (10).

The problem may however be solved by regarding the time as the independent variable. Representing then, as in the last example, the given times by $a, b, \ldots k$, the time sought by x, and the corresponding values of the observed magnitude by $u_a, u_b, \ldots u_k$, and u_x, we must by the solution of the same equation (10) determine x.

The above forms of solution being derived from different hypotheses, will of course differ. We say derived from different hypotheses, because whichsoever element is regarded as dependent is treated not simply as a function, but as a rational and integral function of the other element; and thus the choice affects the nature of the connexion. Except for the avoidance of difficulties of solution, the hypothesis which assumes the time as the independent variable is to be preferred.

Ex. Three observations of a quantity near its time of maximum or minimum being taken, to find its time of maximum or minimum.

Let a, b, c, represent the times of observation, and u_x the magnitude of the quantity at any time x. Then u_a, u_b and u_c are given, and, by Lagrange's formula,

$$u_x = u_a \frac{(x-b)(x-c)}{(a-b)(a-c)} + u_b \frac{(x-c)(x-a)}{(b-c)(b-a)} + u_c \frac{(x-a)(x-b)}{(c-a)(c-b)},$$

and this function of x is to be a maximum or minimum. Hence equating to 0 its differential coefficient with respect to x, we find

$$x = \frac{(b^2-c^2)\,u_a + (c^2-a^2)\,u_b + (a^2-b^2)\,u_c}{2\{(b-c)\,u_a + (c-a)\,u_b + (a-b)\,u_c\}} \qquad (14).$$

This formula enables us to approximate to the meridian altitude of the sun or of a star when a true meridian observation cannot be taken *.

7. As was stated in Art. 4, Lagrange's formula is usually the most convenient for calculating an approximate value of u_x from given observed values of the same when these are not equidistant. But in cases where we have reason to believe that the function is periodic, we may with advantage substitute for it some expression, involving the right number of undetermined coefficients, in which x appears only in the arguments of periodic terms. Thus, if we have $2n+1$ observations, we may assume

$$\begin{aligned} u_x = A_0 + A_1 \cos x + A_2 \cos 2x + \ldots + A_n \cos nx \\ + B_1 \sin x + B_2 \sin 2x + \ldots + B_n \sin nx \qquad (15), \end{aligned}$$

and determine the coefficients by solving the resulting linear equations.

Gauss† has proved that the formula

$$u_x = \frac{\sin\frac{1}{2}(x-b)\sin\frac{1}{2}(x-c)\ldots\sin\frac{1}{2}(x-k)}{\sin\frac{1}{2}(a-b)\sin\frac{1}{2}(a-c)\ldots\sin\frac{1}{2}(a-k)}\,u_a + \ldots \qquad (16),$$

* A special investigation of this problem will be found in Grunert, xxv. 237.

† Werke, Vol. III. p. 281.

is equivalent to (15), $u_a, u_b, \ldots u_k$ being assumed to be the $2n+1$ given values of u_x. It is evident that we obtain $u_x = u_a$ when for x we substitute a in it, and also that when expanded it will only contain sines and cosines of integral multiples of x not greater than nx; and as the coefficients of (15) are fully determinable from the data, it follows that the two expressions are identically equal.

8. Cauchy* has shewn that if $m+n$ values of a function are known, we may find a fraction whose numerator is of the nth, and denominator of the $(m-1)^{\text{th}}$ degree, which will have the same $m+n$ values for the same values of the variable. He gives the general formula for the above fraction, which is somewhat complicated, though obviously satisfying the conditions. We subjoin it for the case when

$$m = 2, \ n = 1,$$

$$u_x = -\frac{u_b u_c (b-c)(x-a) + \ldots}{u_a (b-c)(x-a) + \ldots} \qquad (17).$$

When $m=1$ it reduces of course to Lagrange's formula.

Application to Statistics.

9. When the results of statistical observations are presented in a tabular form it is sometimes required to narrow the intervals to which they correspond, or to fill up some particular hiatus by the interpolation of intermediate values. In applying to this purpose the methods of the foregoing sections, it is not to be forgotten that the assumptions which they involve render our conclusions the less trustworthy in proportion as the matter of inquiry is less under the dominion of any known laws, and that this is still more the case in proportion as the field of observation is too narrow to exhibit fairly the operation of the unknown laws which do exist. The anomalies, for instance, which we meet with in the attempt to estimate the law of human mortality seem rather to

* *Analyse Algébraique*, p. 528, but it is better to read a paper by Brassine (Liouville, XI. 177), in which it is considered more fully and as a case of a more general theorem. This must not be confounded with Cauchy's *Method of Interpolation*, which is of a wholly different character and does not need notice here. He gives it in Liouville, II. 193, and a consideration of the advantages it possesses will be found in a paper by Bienaymé, *Comptes Rendus*, XXXVII. or Liouville, XVIII. 299.

be due to the imperfection of our data than to want of continuity in the law itself. The following is an example of the anomalies in question.

Ex. The expectation of life at a particular age being defined as the average duration of life after that age, it is required from the following data, derived from the Carlisle tables of mortality, to estimate the probable expectation of life at 50 years, and in particular to shew how that estimate is affected by the number of the data taken into account.

Age.	Expectation.	Age.	Expectation.
10	$48{\cdot}82 = u_1$	60	$14{\cdot}34 = u_6$
20	$41{\cdot}46 = u_2$	70	$9{\cdot}18 = u_7$
30	$34{\cdot}34 = u_3$	80	$5{\cdot}51 = u_8$
40	$27{\cdot}61 = u_4$	90	$3{\cdot}28 = u_9$

The expectation of life at 50 would, according to the above scheme, be represented by u_5. Now if we take as our only data the expectation of life at 40 and 60, we find by the method of Art. 3,

$$u_5 = \frac{u_4 + u_6}{2} = 20{\cdot}97 \qquad (a).$$

If we add to our data the expectation at 30 and 70, we find

$$u_5 = \frac{2}{3}(u_4 + u_6) - \frac{1}{6}(u_3 + u_7) = 20{\cdot}71 \qquad (b).$$

If we add the further data for 20 and 80, we find

$$u_5 = \frac{3}{4}(u_4 + u_6) - \frac{3}{10}(u_3 + u_7) + \frac{1}{20}(u_2 + u_8) = 20{\cdot}75 \qquad (c).$$

And if we add in the extreme data for the ages of 10 and 90, we have

$$u_5 = \frac{8}{10}(u_4 + u_6) - \frac{4}{10}(u_3 + u_7)$$

$$+ \frac{8}{70}(u_2 + u_8) - \frac{1}{70}(u_1 + u_9) = 20{\cdot}776 \qquad (d).$$

We notice that the second of the above results is considerably lower than the first, but that the second, third, and fourth exhibit a gradual approximation toward some value not very remote from 20·8.

Nevertheless the actual expectation at 50 as given in the Carlisle tables is 21·11, which is greater than even the first result or the average between the expectations at 40 and 60. We may almost certainly conclude from this that the Carlisle table errs in excess for the age of 50.

And a comparison with some recent tables shews that this is so. From the tables of the Registrar-General, Mr Neison* deduced the following results.

Age.	Expectation.	Age.	Expectation.
10	47·7564	60	14·5854
20	40·6910	70	9·2176
30	34·0990	80	5·2160
40	27·4760	90	2·8930
50	20·8463		

Here the calculated values of the expectation at 50, corresponding to those given in (*a*), (*b*), (*c*), (*d*), will be found to be

21·0307, 20·8215, 20·8464, 20·8454.

We see here that the actual expectation at 50 is less than the mean between those at 40 and 60. We see also that the second result gives a close, and the third a very close, approximation to its value. The deviation in the fourth result, which takes account of the extreme ages of 10 and 90, seems due to the attempt to comprehend under the same law the mortality of childhood and of extreme old age.

When in an extended table of numerical results the differences tend first to diminish and afterwards to increase, and some such disposition has been observed in tables of mortality, it may be concluded that the extreme portions of the tables are subject to different laws. And even should those laws admit, as perhaps they always do, of comprehension under some law higher and more general, it may be inferred that that law is incapable of approximate expression in the particular form (Art. 2) which our methods of interpolation presuppose.

* *Contributions to Vital Statistics*, p. 8.

Areas of Curves.

10. Formulæ of interpolation may be applied to the approximate evaluation of integrals between given limits, and therefore to the determination of the areas of curves, the contents of solids, ... The application is convenient, as it does not require the form of the function under the sign of integration to be known. The process is usually known by the name of Mechanical Quadrature.

PROP. The area of a curve being divided into n portions bounded by $n+1$ equidistant ordinates $u_0, u_1, \dots u_n$, whose values, together with their common distance, are given, an approximate expression for the area is required.

The general expression for an ordinate being u_x, we have, if the common distance of the ordinates be assumed as the unit of measure, to seek an approximate value of $\int_0^n u_x dx$.

Now, by (2),

$$u_x = u_0 + x\Delta u_0 + \frac{x(x-1)}{1.2}\Delta^2 u_0 + \frac{x(x-1)(x-2)}{1.2.3}\Delta^3 u_0 + \dots.$$

Hence

$$\int_0^n u_x dx = u_0 \int_0^n dx + \Delta u_0 \int_0^n x dx + \frac{\Delta^2 u_0}{1.2}\int_0^n x(x-1)\,dx$$
$$+ \frac{\Delta^3 u_0}{1.2.3}\int_0^n x(x-1)(x-2)\,dx + \dots,$$

and effecting the integrations

$$\int_0^n u_x dx = nu_0 + \frac{n^2}{2}\Delta u_0 + \left(\frac{n^3}{3} - \frac{n^2}{2}\right)\frac{\Delta^2 u_0}{1.2} + \left(\frac{n^4}{4} - n^3 + n^2\right)\frac{\Delta^3 u_0}{1.2.3}$$
$$+ \left(\frac{n^5}{5} - \frac{3n^4}{2} + \frac{11n^3}{3} - 3n^2\right)\frac{\Delta^4 u_0}{1.2.3.4}$$
$$+ \left(\frac{n^6}{6} - 2n^5 + \frac{35}{4}n^4 - \frac{50}{3}n^3 + 12n^2\right)\frac{\Delta^5 u_0}{1.2.3.4.5}$$
$$+ \left(\frac{n^7}{7} - \frac{15n^6}{6} + 17n^5 - \frac{225n^4}{4} + \frac{274n^3}{3} - 60n^2\right)\frac{\Delta^6 u_0}{1.2\dots 6}$$
$$+ \dots \qquad (18).$$

It will be observed that the data permit us to calculate the successive differences of u_0 up to $\Delta^n u_0$. Hence, on the assumption that all succeeding differences may be neglected, the above theorem gives an approximate value of the integral sought. The following are particular deductions.

1st. Let $n=2$. Then, rejecting all terms after the one involving $\Delta^2 u_0$, we have

$$\int_0^2 u_x dx = 2u_0 + 2\Delta u_0 + \tfrac{1}{3}\Delta^2 u_0.$$

But $\Delta u_0 = u_1 - u_0$, $\Delta^2 u_0 = u_2 - 2u_1 + u_0$; whence, substituting and reducing,

$$\int_0^2 u_x dx = \frac{u_0 + 4u_1 + u_2}{3}.$$

If the common distance of the ordinates be represented by h, the theorem obviously becomes

$$\int_0^{2h} u_x dx = \frac{u_0 + 4u_h + u_{2h}}{3} h \qquad (19),$$

and is the foundation of a well-known rule in treatises on Mensuration.

2ndly. If there are four ordinates whose common distance is unity, we find in like manner

$$\int_0^3 u_x dx = \frac{3(u_0 + 3u_1 + 3u_2 + u_3)}{8} \qquad (20).$$

3rdly. If five equidistant ordinates are given, we have in like manner

$$\int_0^4 u_x dx = \frac{14(u_0 + u_4) + 64(u_1 + u_3) + 24u_2}{45} \qquad (21).$$

4thly. The supposition that the area is divided into six portions bounded by 7 equidistant ordinates leads to a remarkable result, first given by the late Mr Weddle (*Math. Journal*, Vol. IX. p. 79), and deserves to be considered in detail.

Supposing the common distance of the ordinates to be unity, we find, on making $n=6$ in (18) and calculating the coefficients,

$$\int_0^6 u_x dx = 6u_0 + 18\Delta u_0 + 27\Delta^2 u_0 + 24\Delta^3 u_0 + \frac{123}{10}\Delta^4 u_0$$
$$+ \frac{33}{10}\Delta^5 u_0 + \frac{41}{140}\Delta^6 u_0 \qquad (22).$$

Now the last coefficient $\frac{41}{140}$ differs from $\frac{42}{140}$ or $\frac{3}{10}$ by the small fraction $\frac{1}{140}$, and as from the nature of the approximation we must suppose sixth differences small, since all succeeding differences are to be neglected, we shall commit but a slight error if we change the last term into $\frac{3}{10}\Delta^6 u_0$. Doing this, and then replacing Δu_0 by $u_1 - u_0$ and so on, we find, on reduction,

$$\int_0^6 u_x dx = \frac{3}{10}\{u_0 + u_2 + u_4 + u_6 + 5(u_1 + u_5) + 6u_3\},$$

which, supposing the common distance of the ordinates to be h, gives

$$\int_0^{6h} u_x dx = \frac{3h}{10}\{u_0 + u_{2h} + u_{4h} + u_{6h} + 5(u_h + u_{5h}) + 6u_{3h}\} \qquad (23),$$

the formula required.

It is remarkable that, were the series in the second member of (22) continued, the coefficient of $\Delta^7 u_0$ would be found to vanish. Thus while the above formula gives the exact area when fifth differences are constant, it errs in excess by only $\frac{1}{140}\Delta^6 u_0$ when seventh differences are constant.

The practical rule hence derived, and which ought to find a place in elementary treatises on mensuration, is the following:

The proposed area being divided into six portions by seven equidistant ordinates, add into one sum the even ordinates 5 times the odd ordinates and the middle ordinate, and multiply the result by $\frac{3}{10}$ of the common distance of the ordinates.

Ex. 1. The two radii which form a diameter of a circle are bisected, and perpendicular ordinates are raised at the points of bisection. Required the area of that portion of the circle which is included between the two ordinates, the diameter, and the curve, the radius being supposed equal to unity.

The values of the seven equidistant ordinates are

$$\frac{\sqrt{3}}{2},\ \frac{\sqrt{8}}{3},\ \frac{\sqrt{35}}{6},\ 1,\ \frac{\sqrt{35}}{6},\ \frac{\sqrt{8}}{3},\ \frac{\sqrt{3}}{2},$$

and the common distance of the ordinates is $\frac{1}{6}$. The area hence computed to five places of decimals is ·95661, which, on comparison with the known value $\frac{\pi}{6}+\frac{\sqrt{3}}{4}$, will be found to be correct to the last figure.

The rule for equidistant ordinates commonly employed would give ·95658.

In all these applications it is desirable to avoid extreme differences among the ordinates. Applied to the quadrant of a circle Mr Weddle's rule, though much more accurate than the ordinary one, leads to a result which is correct only to two places of decimals.

Should the function to be integrated become infinite at or within the limits, an appropriate transformation will be needed.

Ex. 2. Required an approximate value of $\int_0^{\frac{\pi}{2}} \log \sin \theta d\theta$.

The function $\log \sin \theta$ becomes infinite at the lower limit. We have, on integrating by parts,

$$\int \log \sin \theta d\theta = \theta \log \sin \theta - \int \theta \cot \theta d\theta,$$

hence, the integrated term vanishing at both limits,

$$\int_0^{\frac{\pi}{2}} \log \sin \theta d\theta = -\int_0^{\frac{\pi}{2}} \theta \cot \theta d\theta.$$

The values of the function $\theta \cot \theta$ being now calculated for the successive values $\theta = 0$, $\theta = \frac{\pi}{12}$, $\theta = \frac{2\pi}{12}$, ... $\theta = \frac{\pi}{2}$, and the theorem being applied, we find

$$-\int_0^{\frac{\pi}{2}} \theta \cot \theta d\theta = -1{\cdot}08873.$$

The true value of the definite integral is known to be

$$\frac{\pi}{2} \log\left(\frac{1}{2}\right), \text{ or } -1{\cdot}08882.$$

11. Lagrange's formula enables us to avoid the intermediate employment of differences, and to calculate directly the coefficient of u_m in the general expression for $\int u_x dx$. If we represent the equidistant ordinates, $2n+1$ in number, by $u_0, u_1 \dots u_{2n}$, and change the origin of the integrations by assuming $x - n = y$, we find ultimately

$$\int_0^{2n} u_x dx = A_0 u_n + A_1 (u_{n+1} + u_{n-1}) + A_2 (u_{n+2} + u_{n-2}) \dots + A_n (u_{2n} + u_0),$$

where generally

$$A_r = \frac{(-1)^r}{1.2\dots(n+r)\,1.2\dots(n-r)} \times \int_{-n}^{n} \frac{(0^2 - y^2)(1^2 - y^2)\dots(n^2 - y^2)}{r^2 - y^2} dy \qquad (24).$$

A similar formula may be established when the number of equidistant ordinates is even.

12. The above method of finding an approximate value for the area of a curve between given limits is due to Newton and Cotes. It consists in expressing this area in terms of observed values of equidistant ordinates in the form

$$\text{Area} = A_0 u_0 + A_1 u_1 + \dots,$$

where A_0, A_1, ... are coefficients *depending solely on the number of ordinates observed*, and thus calculable beforehand and the same for all forms of u_x. It is however by no means necessary that the ordinates should be equidistant; Lagrange's formula enables us to express the area in terms of any n ordinates, and gives

$$\int_q^p u_x dx = A_a u_a + A_b u_b + \dots \qquad (25),$$

where
$$A_a = \int_q^p \frac{(x-b)(x-c)\dots}{(a-b)(a-c)\dots} dx \qquad (26).$$

Now it is evident that the closeness of the approximation depends, first, on the number of ordinates observed, and secondly, on the nature of the function u_x. If, for instance, u_x be a rational integral function of x of a degree not higher than the $(n-1)^{\text{th}}$, the function is fully determined when n ordinates are given, whether these be equidistant or not, and the above formula gives the area exactly.

If this be not the case, it is evident that different sets of observed ordinates will give different values for the area, the difference between such values measuring the degree of the approximation. Some of these will be nearer to the actual value than others, but it would seem probable that a knowledge of the form of u_x would be required to enable us to choose the best system. But Gauss* has demonstrated that we can, without any such knowledge, render our approximation accurate when u_x is of a degree not higher than the $(2n-1)^{\text{th}}$ if we choose rightly the position of the n observed ordinates.

This amounts to doubling the degree of the approximation, so that we can find accurately the area of the curve $y = u_x$ between the ordinates to $x = p$, $x = q$, by observing n properly chosen ordinates, although u_x be of the $(2n-1)^{\text{th}}$ degree.

The following proof of this most remarkable proposition is substantially the same as that given by Jacobi (*Crelle*, Vol. I. 301).

* Werke, Vol. III. p. 203.

Let $\int_q^p u_x dx$ be the integral whose value is required, where u_x is a rational and integral function of the $(2n-1)^{\text{th}}$ degree. Let $u_a, u_b \dots$ be the n observed ordinates, and $f(x)$ the expression which they give for u_x by substitution in Lagrange's formula. Let

$$A(x-a)(x-b) \dots \equiv M,$$

where A is a constant.

Since $u_x - f(x)$ vanishes when $x = a, b, \dots$ it must be equal to MN where N is rational, integral, and of the $(n-1)^{\text{th}}$ degree, and the error in the approximation is $\int_q^p MNdx$, which we shall now shew can be made to vanish by properly choosing M, i.e. by properly choosing the ordinates measured.

Now

$$\int MN\,dx = M_1N - \int M_1N'\,dx$$

$$= M_1N - M_2N' + \int M_2N''\,dx = \&\text{c.}$$

$$= M_1N - M_2N' + \&\text{c.} - (-1)^n M_nN^{(n-1)},$$

denoting by M_κ the result of integrating M κ times, and by $N^{(\kappa)}$ the result of differentiating N κ times; and remembering that $N^{(n-1)}$ is a constant.

Taking the above integrals between the given limits, we see that the problem reduces to making M_r vanish at each limit for all values of r from $r = 1$ to $r = n$.

This is at once accomplished by taking

$$M \equiv \frac{d^n\,\{(x-p)(x-q)\}^n}{dx^n},$$

for it is thus a rational and integral function of x of the nth degree, such that all its first n integrals can be taken to vanish at the given limits. That this is the case is seen at once when we consider that the parts independent of the arbitrary constants will contain some power of $(x-p)(x-q)$ as a factor, and will thus vanish at both limits.

The coefficients $A_a, A_b \dots$ in $\int f(x)\,dx$ will of course be functions of p and q of the form given in (26). In order to save the trouble of calculating them for all values of the limits, it is usual to transform the integral, previously to applying the above theorem, so as to make the limits 1 and -1. We then have

$$M \equiv \frac{d^n (x^2-1)^n}{dx^n} \equiv \frac{\lfloor 2n}{\lfloor n} \left\{ x^n - \frac{n^2 (n-1)}{2n(2n-1)} x^{n-2} \right.$$

$$\left. + \frac{n^2 (n-1)^2 (n-2)(n-3)}{1.2.2n(2n-1)(2n-2)(2n-3)} x^{n-4} - \dots \right\},$$

and $a, b, c \dots$ are the roots of $M=0$, which are known to be real, since those of $(x^2-1)^n = 0$ are all real.

13. We shall now proceed to demonstrate a most important formula for the mechanical quadrature of curves. It was first given by Laplace*, and will be seen to be closely allied to (18).

Since

$$1+\Delta = \epsilon^D, \therefore \Delta \equiv D\frac{\Delta}{D} = D\left\{\frac{\Delta}{\log(1+\Delta)}\right\}$$

$$\Delta w_x = \frac{d}{dx}\left\{\frac{\Delta}{\log(1+\Delta)}\right\} w_x$$

$$= \frac{d}{dx}\left\{1+\frac{\Delta}{2}-\frac{1}{12}\Delta^2+\frac{1}{24}\Delta^3-\dots\right\} w_x \dagger.$$

Integrate between limits 1 and 0, remembering that

$$[w_x]_{x=r}^{x=r+1} = \Delta w_r,$$

* *Mécanique Céleste*, IV. 207.

† The coefficients of the powers of t in $\frac{t}{\log(1+t)}$ may be calculated either directly, or by the method in Ex. 18 at the end of this Chapter.

and we easily get, writing u_x for Δw_x,

$$\int_0^1 u_x dx = \left\{1 + \frac{\Delta}{2} - \dots\right\} u_0$$

$$= \frac{u_0 + u_1}{2} - \frac{1}{12}\Delta^2 u_0 + \frac{1}{24}\Delta^3 u_0 - \dots.$$

Writing down similar expressions for $\int_1^2 u_x dx$, ... and adding, we obtain

$$\int_0^n u_x dx = \frac{u_0}{2} + u_1 + u_2 + \dots + \frac{u_n}{2}$$

$$- \frac{1}{12}(\Delta u_n - \Delta u_0)$$

$$+ \frac{1}{24}(\Delta^2 u_n - \Delta^2 u_0)$$

$$- \dots \qquad (27),$$

since

$$\Delta^r u_0 + \Delta^r u_1 + \dots = \Delta^{r-1}(\Delta u_0 + \Delta u_1 + \dots) = \Delta^{r-1}(u_n - u_0).$$

This formula has the disadvantage of containing the differences of u_n, which cannot be calculated from the values $u_0, u_1 \dots u_n$. We may remedy this in the following way:

$$\frac{\Delta}{\log(1+\Delta)} = E \frac{\dfrac{-\Delta}{1+\Delta}}{\log\left(1 - \dfrac{\Delta}{1+\Delta}\right)} = E \frac{-\Delta E^{-1}}{\log(1 - \Delta E^{-1})};$$

$$\therefore \left(1 + \frac{\Delta}{2} - \frac{1}{12}\Delta^2 + \frac{1}{24}\Delta^3 - \dots\right) w_n$$

$$= E\left\{1 - \frac{1}{2}\Delta E^{-1} - \frac{1}{12}\Delta^2 E^{-2} - \frac{1}{24}\Delta^3 E^{-3} - \dots\right\} w_n.$$

Removing the first two terms from each side since they are obviously equal, and writing u_n for Δw_n, we get

$$-\frac{1}{12}\Delta u_n + \frac{1}{24}\Delta^2 u_n - \ldots = -\frac{1}{12}\Delta u_{n-1} - \frac{1}{24}\Delta^2 u_{n-2} - \ldots$$

and the formula becomes

$$\int_0^n u_x dx = \frac{u_0}{2} + u_1 + u_2 + \ldots + \frac{u_n}{2}$$
$$-\frac{1}{12}(\Delta u_{n-1} - \Delta u_0)$$
$$-\frac{1}{24}(\Delta^2 u_{n-2} + \Delta^2 u_0)$$
$$- \ldots \qquad (28).$$

In the above investigation we have in reality twice performed the operation $\frac{1}{\Delta}$ on both sides of an equation. We shall see that $\Delta u_x = \Delta v_x$ only enables us to say $u_x = v_x + C$ and not $u_x = v_x$; hence we should have added an arbitrary constant. But the slightest consideration is sufficient to shew that this constant will in each case be zero.

14. The problems of Interpolation and Mechanical Quadrature are of the greatest practical importance, the formulæ deduced therefrom being used in all extended calculations in order to shorten the labour without affecting greatly the accuracy of the result. This they are well capable of doing; indeed Olivier maintains (*Crelle*, II. 252) that calculations proceeding by Differences will probably give a closer approximation to the exact result than corresponding ones that proceed by Differential Coefficients. In consequence of this practical value many Interpolation-formulæ have been arrived at by mathematicians who have had to do with actual calculations, each being particularly suited to some particular calculation. All the most celebrated of these formulæ will be found in the accompanying examples. Examples of calculations based upon them can usually be found through the references; the papers by Grunert (*Archiv*, XIV. 225 and XX. 361), which contain a full inquiry into the subject, may also be consulted for this purpose. Numerical examples of the application of several Interpolation-formulæ may also be found in a paper by Hansen (*Relationen zwischen Summen und Differenzen, Abhandlungen der Kön. Sächs. Gesellschaft*, 1865), in which also he gives a very detailed inquiry into the various methods in use, with numerical calculation of coefficients, ... We must warn the reader against the notation, which is unscientific and wholly in defiance of convention, e.g. $\Delta y_{x+\frac{1}{2}}$ and

$\Delta^2 y_x$ are used to represent the Δy_x and $\Delta^2 y_{x-1}$ of the ordinary notation. A good paper on the subject by Encke (*Berlin. Astron. Jahrbuch*, 1830), from which Ex. 7 is taken, labours under the same disadvantage; and Stirling's formula (Ex. 9) is seldom found stated in the correct notation.

In speaking of the developments which the theory has received we must mention an important *Mémoire* by Jacobi (*Crelle*, XXX. 127) on the Cauchy Interpolation-formula of Art. 8. In it the author points out the advantages that it possesses over others, and subjects it to a very full investigation, representing the numerator and denominator in various forms as determinants, and considering especially the case when two or more of the values of the independent variable approach equality. A paper by Rosenhain which follows immediately after it treats also of the above formula in representing the condition that two equations $\phi(x)=0$ and $f(x)=0$ should have a common root, in terms of the values of the expression $\dfrac{\phi(x)}{f(x)}$ for different values of x.

But the most important researches in the theory of Interpolation have had reference to the Gauss-formula of Art. 12. Minding (*Crelle*, VI. 91) extends it to the approximate evaluation of double integrals between constant limits. Christoffel (*Crelle*, LV. 61) investigates the more general problem of determining the ordinates we should choose for observation when certain ordinates are already given, so that the approximation may be as close as possible. Mehler (*Crelle*, LXIII. 152) shews that a closely analogous method enables us to calculate integrals of the form

$$\int_{-1}^{1} (1-x)^{\lambda} (1+x)^{\mu} f(x)\, dx$$

with great accuracy, the position of the ordinates chosen being in this case determined by the roots of the equation of the nth degree

$$(1-x)^{-\lambda}(1+x)^{-\mu}\frac{d^n}{dx^n}\left\{(1-x)^{n+\lambda}(1+x)^{n+\mu}\right\}=0,$$

λ and μ being each > -1.

Jacobi had previously examined the case in which $\lambda=\mu=-\dfrac{1}{2}$; in other words, he had shewn that in

$$\int_{-1}^{1}\frac{f(x)}{\sqrt{1-x^2}}\,dx \text{ or } \int_0^{\pi} f(\cos\theta)\,d\theta,$$

the positions of the co-ordinates to be chosen after the analogy of the Gauss-formula are given by the roots of

$$\sqrt{1-x^2}\,\frac{d^n}{dx^n}(1-x^2)^{n-\frac{1}{2}}=0,$$

which is equivalent to $\cos(n\cos^{-1}x)=0$. Hence $x=\cos\dfrac{2m+1}{2n}\pi$.

In this case the coefficients A_a, A_b, ... (see (26), page 51) are all equal, each being $\dfrac{\pi}{n}$, and the formula becomes

$$\int_0^{\pi} f(\cos\theta)\,d\theta=\frac{\pi}{n}\left\{f\left(\cos\frac{\pi}{2n}\right)+f\left(\cos\frac{3\pi}{2n}\right)+\ldots+f\left(\cos\frac{2n-1}{2n}\pi\right)\right\}.$$

In most of the above papers the magnitude of the error caused by using the approximate formula instead of the exact value of the function is investigated.

The special importance of the method becomes evident when we consider the close relation between it and the celebrated Laplace's functions. This is seen by comparing the expression for the nth Laplace's coefficient of one variable,

$$P_n = \frac{1}{2^n \underline{|n}} \cdot \frac{d^n (x^2-1)^n}{dx^n},$$

with the value of M in Art. 12; and the similarity of the corresponding expressions for two variables is equally great. In fact the Gauss-method may be represented as follows:—

Let u_x be a rational integral function of the $(2n-1)^{\text{th}}$ degree, and Y_n be the nth Laplace's coefficient. Divide u_x by Y_n, and let N be the quotient and $f(x)$ the remainder which is of the $(n-1)^{\text{th}}$ degree. Thus $u_x = f(x) + Y_n \cdot N$. Integrate between the limits 1 and -1, and since N is of a lower degree than Y_n, $\int_{-1}^{1} Y_n N dx = 0$, and we are left with $\int_{-1}^{1} f(x)\, dx$ which is accurately found by the Lagrange-formula from the n observed values of u_x.

In consequence of this close connexion the method is of great importance in the investigation of Laplace's Functions and of the kindred subject of Hypergeometrical Series. Heine's *Handbuch der Kugelfunctionen* will supply the reader with materials for discovering the exact relation in which they stand to one another, or he may compare a paper by Bauer on Laplace's functions (*Crelle*, LVI. 101) with that by Christoffel given above. For instances of numerical calculation he may consult Bertrand (*Int. Cal.* 339), where, however, the limits 1 and 0 are taken.

EXERCISES.

1. Required, an approximate value of log 212 from the following data:

$$\log 210 = 2{\cdot}3222193, \qquad \log 213 = 2{\cdot}3283796,$$
$$\log 211 = 2{\cdot}3242825, \qquad \log 214 = 2{\cdot}3304138.$$

2. Find a rational and integral function of x of as low a degree as possible that shall assume the values 3, 12, 15, and -21, when x is equal to 3, 2, 1, and -1 respectively.

3. Express v_2 and v_3 approximately, in terms of v_0, v_1, v_4, and v_5, both by Lagrange's formula and the method of (7), Art. 4.

4. The logarithms in Tables of n decimal places differ from the true values by $\pm\frac{5}{10^{n+1}}$ at most. Hence shew that the errors of logarithms of n places obtained from the Tables by interpolating to first and second differences cannot exceed $\pm\frac{1}{10^n}+e$ and $\pm\frac{1}{10^n}\times\frac{9}{8}+e'$ respectively, e and e' being the errors due exclusively to interpolation. (Smith's *Prize.*)

5. The values of a function of the time are a_1, a_2, a_3, a_4, at epochs separated by the common interval h; the first differences are d_1, d'_1, d''_1, the second differences are d_2, d'_2, and the third difference d_3. Hence obtain the following formulæ of interpolation to third differences:

$$f(t)=a_2+\left(d'_1-\frac{d_2}{2}-\frac{d_3}{6}\right)\frac{t}{h}+\frac{d_2}{2}\cdot\frac{t^2}{h^2}+\frac{d_3}{6}\cdot\frac{t^3}{h^3};$$

or
$$f(t)=a_3+\left(d'_1+\frac{d'_2}{2}-\frac{d_3}{6}\right)\frac{t}{h}+\frac{d'_2}{2}\cdot\frac{t^2}{h^2}+\frac{d_3}{6}\cdot\frac{t^3}{h^3};$$

t being reckoned in the first case from the epoch of a_2, and in the second from that of a_3.

6. If $P, Q, R, S, \ldots$ be the values of X, an unknown function of x, corresponding to $x=p, q, r, s, \ldots$, shew that (under the same hypothesis as in the case of Lagrange's formula),

$$X=P+(x-p)\{p,q\}+(x-p)(x-q)\{p,q,r\}+\ldots,$$

where generally

$$\{p,q,r\ldots\}\equiv\frac{P}{(p-q)(p-r)\ldots}+\frac{Q}{(q-p)(q-r)\ldots}+\ldots.$$

7. Shew that, in the notation of the last question, if $q-p=r-q=s-r=\ldots=1$,

$$\{p,q,r,s\}=\frac{\Delta^3P}{1.2.3};$$

and apply the theorem to demonstrate that

(1) $$u_{n+x} = u_n + x\Delta u_n + \frac{x(x-1)}{1.2}\Delta^2 u_{n-1} + \frac{x(x-1)(x+1)}{1.2.3}\Delta^3 u_{n-1} + \frac{x(x^2-1)(x-2)}{1.2.3.4}\Delta^4 u_{n-2} + \dots$$

(2) $$u_{n+x} = u_n + x\Delta u_{n-1} + \frac{x(x+1)}{1.2}\Delta^2 u_{n-1} + \frac{x(x-1)(x+1)}{1.2.3}\Delta^3 u_{n-2} + \frac{x(x^2-1)(x+2)}{1.2.3.4}\Delta^4 u_{n-2} + \dots$$

8. Shew that the function

$$1 + \frac{t-a}{a-b} + \frac{(t-a)(t-b)}{(a-b)(a-c)} + \dots$$

becomes unity when $t = a$, and zero when $t = b, c, \dots$, and deduce Ex. 6 therefrom.

9. Demonstrate Stirling's Interpolation-formula

$$u_t = u_0 + \frac{t}{2}\Delta(u_0 + u_{-1}) + \frac{t^2}{1.2}\Delta^2 u_{-1} + \frac{t(t^2-1)}{2.1.2.3}\Delta^3(u_{-1} + u_{-2}) + \frac{t^2(t^2-1)}{1.2.3.4}\Delta^4 u_{-2} + \dots$$

(Smith's *Prize*, 1860.)

10. Deduce Newton's formula for Interpolation from Lagrange's when the values are equidistant.

11. If μ radii vectores (μ being an odd integer) be drawn from the pole dividing the four right angles into equal parts, shew that an approximate value of a radius vector (u_θ) which makes an angle θ with the initial line is

$$u_\theta = \frac{1}{\mu}\Sigma \frac{\sin\frac{\mu}{2}(\theta - a)}{\sin\frac{1}{2}(\theta - a)} u_a,$$

where $a, b, \dots$ are the angles that the μ radii vectores make with the initial line.

12. Assuming the formula for resolving

$$\frac{f(x)}{(x-a)(x-b)\dots(x-k)}$$

into Partial Fractions, deduce Lagrange's Interpolation-formula.

13. If $\phi(x)=0$ be a rational algebraical equation in x of any order, and $z_1, z_2 \dots z_k$ be taken to represent $\phi(1)$, $\phi(2), \dots \phi(k)$, find under what conditions

$$z_1 z_2 \dots z_k \sum_{r=1}^{r=k} \frac{r}{z_r(z_1-z_r)\dots(z_k-z_r)}$$

may be taken as an approximate root of the equation.

14. Demonstrate Simpson's rule for finding an approximate value for the area of a curve, when an odd number of equidistant ordinates are known, viz.: To four times the sum of the even ordinates add twice the sum of the odd ones; subtract the sum of the extreme ordinates and multiply the result by one-third the common distance.

15*. Shew that Simpson's rule is tantamount to considering the curve between two consecutive odd ordinates as parabolic. Also, if we assume that the curve between each ordinate is parabolic, and that it also passes through the extremity of the next ordinate (the axes of the parabolæ being in all cases parallel to the axis of y), the area will be given by

$$\text{Area} = h\left[\Sigma y - \frac{1}{24}\left\{15(y_0+y_n) - 4(y_1+y_{n-1}) + \overline{y_2+y_{n-2}}\right\}\right].$$

16†. Given u_x and u_{x+1}, and their even distances, shew that

$$u_{x+\frac{1}{2}} = \frac{1}{2}\left\{1 - \frac{1}{8}\Delta^2 + \frac{1.3}{8.16}\Delta^4 - \frac{1.3.5}{8.16.24}\Delta^6 + \dots\right\}\overline{u_x + u_{x+1}}.$$

* On the comparative merits of these and similar methods see Dupain (*Nouvelles Annales*, XVII. 288).

† The notation in this formula (due to Gauss) is that referred to on the top of page 56.

17. Shew that

$$u_{n+x} = u_n + x\Delta u_{n-r} + \frac{x(x+2r-1)}{1.2}\Delta^2 u_{n-2r} + \frac{x(x+3r-1)(x+3r-2)}{1.2.3}\Delta^3 u_{n-3r} + \dots$$

$$\Delta^n u_x = \Delta^n u_{x-n} + n\Delta^{n+1} u_{x-n-1} + \frac{n(n+1)}{1.2}\Delta^{n+2} u_{x-n-2} + \dots$$

In what cases would the above formulæ be especially useful?

18. Shew that the coefficient of $\Delta^r u_n$ in (27) is equal to

$$\int_0^1 \frac{x^{(r+1)}}{\underline{|r+1}} dx,$$

and hence shew the exact relationship in which (27) and (18) stand to each other.

19*. If from the values u_a, u_b ... of a function corresponding to values a, b, c ... of the variable, we obtain an Interpolation-formula,

$$u_x = u_a + B(x-a) + C(x-a)(x-b) + D(x-a)(x-b)(x-c) + \dots,$$

shew that

$$B = \frac{\Delta u_a}{b-a}, \quad C = \frac{\Delta B}{c-a}, \quad D = \frac{\Delta C}{d-a}, \quad \dots$$

where $\Delta\phi(a, b, \dots) \equiv \phi(b, c, \dots) - \phi(a, b, \dots)$.

Deduce (2), page 35, from the above formula.

* Newton's *Principia*, Lemma v. Lib. III. This is the first attempt at finding a general Interpolation-formula, and gives a complete solution of the problem. The result is of course identically that obtained by Lagrange's formula, though in a very different form.

CHAPTER IV.

FINITE INTEGRATION, AND THE SUMMATION OF SERIES.

1. THE term integration is here used to denote the process by which, from a given proposed function of x, we determine some other function of which the given function expresses the *difference.*

Thus to integrate u_x is to find a function v_x such that

$$\Delta v_x = u_x.$$

The operation of integration is therefore by definition the inverse of the operation denoted by the symbol Δ. As such, it may with perfect propriety be denoted by the inverse form Δ^{-1}. It is usual however to employ for this purpose a distinct symbol, Σ, the origin of which, as well as of the term integration by which its office is denoted, it will be proper to explain.

One of the most important applications of the Calculus of Finite Differences is to the finite summation of series.

Now let $u_0, u_1, u_2, \ldots$ represent successive terms of a series whose general term is u_x, and let

$$v_x = u_a + u_{a+1} + u_{a+2} \ldots + u_{x-1} \qquad (1).$$

Then, a being constant so that u_a remains the initial term, we have

$$v_{x+1} = u_a + u_{a+1} + \ldots + u_{x-1} + u_x. \qquad (2).$$

Hence, subtracting (1) from (2),

$$\Delta v_x = u_x, \quad \therefore \quad v_x = \Delta^{-1} u_x.$$

It appears from the last equation that Δ^{-1} applied to u_x expresses the sum of that portion of a series whose general term is u_x, which begins with a *fixed* term u_a and ends with u_{x-1}. On this account Δ^{-1} has been usually replaced by the symbol Σ, considered as indicating a *summation* or *integration.* At the same time the properties of the symbol Σ, and the mode of performing the operation which it denotes, or, to speak with greater strictness, of answering that question of which it is virtually an expression, are best deduced, and are usually deduced, from its definition as the inverse of the symbol Δ.

Now if we consider Σu_x as defined by the equation

$$\Sigma u_x = u_{x-1} + u_{x-2} + \ldots + u_a. \qquad (3),$$

it denotes a direct and always possible operation, but if we consider it as defined by the equation

$$\Sigma u_x = \Delta^{-1} u_x \qquad (4),$$

and as having for its object the discovery of some finite expression v_x, which satisfies the equation $\Delta v_x = u_x$, it is interrogative rather than directive (*Diff. Equat.* p. 376, 1st ed.), it sets before us an object of enquiry but does not prescribe any mode of arriving at that object; nor does it give us the assurance that there is but one answer to the question it virtually propounds. A moment's consideration, indeed, will assure us that the number of expressions that can claim to be denoted by $\Delta^{-1}u_x$ is infinite, since it includes the quantity

$$u_a + u_{a+1} + \ldots + u_{x-1},$$

whatever value a may be supposed to have, provided only that it is one of the series of integral values which x is supposed to take. We cannot therefore consider the definitions of Σu_x contained in (3) and (4) as identical, and shall therefore proceed to investigate the relation between them and the restrictions as to the use of each.

It is obvious that the Σu_x of (3) is one of the functions represented by the $\Delta^{-1}u_x$ in (4), since it satisfies the equation $\Delta v_x = u_x$. But this is of no value to us unless we can recognize to which of the functions represented by $\Delta^{-1}u_x$ in (4) it is equal, or obtain an expression for it in terms of any one of them. This last we shall now proceed to do.

Let $\phi(x)$ be a function such that $\Delta\phi(x) = u_x$.

$$\therefore \phi(a+1) - \phi(a) = u_a,$$
$$\phi(a+2) - \phi(a+1) = u_{a+1},$$
$$\ldots\ldots\ldots\ldots\ldots\ldots\ldots\ldots\ldots\ldots$$
$$\phi(x) - \phi(x-1) = u_{x-1},$$
$$\therefore \phi(x) - \phi(a) = u_a + u_{a+1} \ldots\ldots + u_{x-1} = \Sigma u_x \text{ in (3)}.$$

Hence retaining for Σu_x the definition of (4) we should write (3) thus:

$$^*\Sigma u_x - \Sigma u_a = u_a + u_{a+1} \ldots\ldots + u_{x-1} \qquad (5).$$

Again suppose Σu_x to be defined by (3) and be equal to $\phi(x)$, and let the Σu_x of (4) be given generally by $\phi(x) + w_x$,

$$\text{then } u_x = \Delta\{\phi(x) + w_x\} = \Delta\phi(x) + \Delta w_x = u_x + \Delta w_x;$$

$\therefore \Delta w_x = 0$, or w_x does not change when x is increased by unity; hence it remains constant while x takes all the series of values which it is permitted to take in any problem in Finite Differences. Since then w_x will remain unchanged, so far as we shall have to do with it, we shall denote it by C and regard it as a constant, and examine its true nature later on. (Art. 4, Ch. II.)

Hence regarding Σu_x as defined by (3) we should write (4) thus:

$$\Delta^{-1} u_x = \Sigma u_x + C \qquad (6).$$

* Were it not that in so fundamental a theorem it is advisable to use only such methods as are beyond all suspicion as to their rigour, we might have arrived more easily at the same result symbolically, thus:

$$u_a + u_{a+1} + \ldots + u_{x-1} = \{1 + E + E^2 + \ldots + E^{x-a-1}\} u_a$$
$$= \frac{E^{x-a} - 1}{E - 1} u_a = (E^{x-a} - 1)\,\Delta^{-1} u_a = (E^{x-a} - 1)\,\Sigma u_a, \text{ from (4)} \ldots (7),$$
$$= \Sigma u_x - \Sigma u_a \qquad (8),$$

which agrees with (5). But the method in the text is preferable, since the steps in (7) and (8) presuppose a rigorous examination into the nature of the symbols Δ^{-1} and Σ before we can state the arithmetical equivalence of the quantities with which we are dealing, i.e. some such investigation as that in the text.

We shall not dwell farther on this point, since the difference between the Σu_x of (3) and that of (4) is precisely analogous to that between the definite integral $\int_a^b \phi(x)\,dx$, and the indefinite integral $\int \phi(x)\,dx$, and the precautions necessary to be taken in using them are identical with those to which we are accustomed in the Integral Calculus. In fact we adopt a notation for *definite* Finite Integrals strikingly similar to that for Definite Integrals in the Infinitesimal Calculus, writing the Σu_x of (3) in the form $\sum_{r=a}^{r=x-1} u_r$.

Integrable Forms.

2. As in Integral Calculus, we shall be able to obtain finite expressions for the integrals of but few forms, and must be content to express the integrals of others in the form of infinite series. Of such integrable forms the following are the most important, as being of frequent recurrence and reducible under general laws.

1st Form. Factorial expressions of the form

$$x(x-1)\ldots(x-m+1) \text{ or } x^{(m)}$$

in the notation of Ch. II. Art. 2.

We have

$$\Delta x^{(m+1)} = (m+1)\,x^{(m)};$$

$$\therefore\ \Sigma x^{(m)} = \frac{x^{(m+1)}}{m+1} + C,$$

or $$\Sigma x(x-1)\ldots(x-m+1) = \frac{x(x-1)\ldots(x-m)}{m+1} + C \qquad (1).$$

Taking this between limits $x=n$ and $x=m$, $(n>m)$, we get

$$1.2\ldots m + 2.3\ldots(m+1) + \ldots + (n-m)\ldots(n-2)(n-1)$$
$$= \frac{n(n-1)\ldots(n-m)}{m+1}.$$

Or we may retain C and determine it subsequently, thus

$$1.2...m+2.3...(m+1)+...+(n-m)...(n-2)(n-1)$$
$$=\frac{n(n-1)...(n-m)}{m+1}+C.$$

Put $n=m+1$ and the series on the left-hand side reduces to its first term, and we obtain

$$1.2...m=\frac{(m+1)m...1}{m+1}+C;\quad \therefore\ C=0.$$

Thus also if $u_x \equiv ax+b$, we have

$$\Sigma u_x u_{x-1}...u_{x-m+1}=\frac{u_x u_{x-1}...u_{x-m}}{a(m+1)}+C \qquad (2).$$

Ex. 1. Sum the series

$$3.5.7+5.7.9+... \text{ to } n \text{ terms.}$$

Here $a=2$, $b=5$, $m=3$, and since we have to find the sum of n terms we must change n into $n+1$ in the last formula, and we obtain

$$\Sigma(2n+7)(2n+5)(2n+3)$$
$$=\frac{(2n+7)(2n+5)(2n+3)(2n+1)}{4\times 2}+C.$$

But $n=1$ gives us

$$3.5.7=\frac{9\times 7\times 5\times 3}{4.2}+C;\quad \therefore\ C=-\frac{105}{8};$$

$$\therefore\ 3.5.7+5.7.9+... \text{ to } n \text{ terms}$$
$$=\frac{(2n+7)(2n+5)(2n+3)(2n+1)}{8}-\frac{105}{8}.$$

2nd Form. Factorial expressions of the form

$$\frac{1}{x(x+1)...(x+m-1)} \text{ or } x^{(-m)}.$$

We have by Ch. II. Art. 2,

$$\Delta x^{(-m+1)} = (-m+1)\, x^{(-m)};$$

$$\therefore\ \Sigma x^{(-m)} = \frac{x^{(-m+1)}}{-m+1} + C \qquad (3).$$

So also if $u_x \equiv ax + b$, we have

$$\Delta \frac{1}{u_x u_{x+1} \dots u_{x+m-1}} = \frac{-am}{u_x u_{x+1} \dots u_{x+m}} \qquad (4);$$

$$\therefore\ \Sigma \frac{1}{u_x u_{x+1} \dots u_{x+m}} = C - \frac{1}{amu_x u_{x+1} \dots u_{x+m-1}};$$

or, writing $m-1$ for m,

$$\Sigma \frac{1}{u_x u_{x+1} \dots u_{x+m-1}} = C - \frac{1}{a(m-1)\, u_x u_{x+1} \dots u_{x+m-2}} \qquad (5).$$

It will be observed that there must be at least two factors in the denominator of the expression to be integrated. No finite expression exists for $\Sigma \dfrac{1}{ax+b}$.

Ex. 2. Find the sum of n terms of the series

$$\frac{1}{1.4.7} + \frac{1}{4.7.10} + \dots .$$

We have here $a=3$, $b=-2$, $m=3$.

$\therefore$ Sum of $(n-1)$ terms

$$= \Sigma \frac{1}{u_n u_{n+1} u_{n+2}} = C - \frac{1}{3 \times 2 \times u_n . u_{n+1}}$$

$$= C - \frac{1}{6\,(3n-2)\,(3n+1)}.$$

Put $n=2$ and we obtain

$$\frac{1}{1.4.7} = C - \frac{1}{6.4.7};\ \therefore\ C = \frac{1}{24}.$$

Hence (writing n for $n-1$ and therefore $n+1$ for n)

$$\text{Sum of } n \text{ terms} = \frac{1}{24} - \frac{1}{6(3n+1)(3n+4)}.$$

As all that is known of the integration of rational functions is virtually contained in the two primary theorems of (2) and (5), it is desirable to express these in the simplest form*. Supposing then $u_x \equiv ax+b$, let

$$u_x u_{x-1} \dots u_{x-m+1} \equiv (ax+b)^{(m)},$$

$$\frac{1}{u_x u_{x+1} \dots u_{x+m-1}} \equiv (ax+b)^{(-m)},$$

then

$$\Sigma\,(ax+b)^{(m)} = \frac{(ax+b)^{(m+1)}}{a(m+1)} + C \qquad (6),$$

whether m be positive or negative. The analogy of this result with the theorem

$$\int (ax+b)^m dx = \frac{(ax+b)^{m+1}}{a(m+1)} + C$$

is obvious.

We shall now shew how to reduce other forms to one of the preceding.

3rd Form. Rational and integral functions.

* As most of the summations of series whose n^{th} term is a rational function of n will have to be effected by these methods, and as such summations are of very frequent occurrence, it is still more important to have a readily applicable rule for effecting them. The following is perhaps the most convenient form for finding the sum of n terms of such series:—

"Write down the nth term with its factors in ascending order of magnitude, $\left\{\begin{array}{l}\text{add one factor at the end}\\\text{take away one factor at the beginning}\end{array}\right\}$, divide by the number of factors now remaining, and by the coefficient of x (in each factor), and $\left\{\begin{array}{l}\text{add to}\\\text{subtract from}\end{array}\right\}$ a constant."

It is scarcely necessary to add that the upper line in the brackets must be taken when the terms are of the form $u_x u_{x-1} \dots u_{x-m+1}$ and the lower when of the form $\dfrac{1}{u_x u_{x+1} \dots u_{x+m-1}}$.

By Ch. II. Art. 5

$$\phi(x) = \phi(0) + \Delta\phi(0)\,x + \frac{\Delta^2\phi(0)}{1.2}x^{(2)} + \dots \qquad (6').$$

Let $\phi(x) \equiv \Sigma v_x$ and put C for $\phi(0)$,

$$\therefore\ \Sigma v_x = C + x\,.\,v_0 + \frac{x^{(2)}}{1\,.\,2}\,.\,\Delta v_0 + \dots \qquad (7),$$

and the number of terms will be finite if v_x be rational and integral.

The series in (6') comes from the equivalence of the operations denoted by the symbols E^x and $(1+\Delta)^x$. In like manner we may obtain a cognate expression from the equivalence of E^{-x} and $(1+\Delta)^{-x}$. This gives us, when we perform them on $\phi(x)$,

$$\phi(0) = \phi(x) - x\,.\,\Delta\phi(x) + \frac{x(x+1)}{1.2}\Delta^2\phi(x) - \dots.$$

Putting as before $\phi(x) \equiv \Sigma v_x$ and C for $\phi(0)$, and transposing, we get

$$\Sigma v_x = C + xv_x - \frac{x(x+1)}{1.2}\Delta v_x + \dots \qquad (8)^*.$$

In applying the above to the summation of series we may avoid the use of an undetermined constant and render the demonstration more direct by proceeding as follows:

$$v_a + v_{a+1} + \dots + v_{a+x-1} = \{1 + E + E^2 + \dots E^{x-1}\}v_a$$

$$= \frac{E^x - 1}{E - 1}v_a = \frac{(1+\Delta)^x - 1}{\Delta}v_a$$

$$= \left\{x + \frac{x(x-1)}{1.2}\Delta + \dots\right\}v_a \qquad (9).$$

*. That the constants in (7) and (8) are the same appears evident when we consider that (8) may be obtained from (7) by mere algebraical transformation. The series-portions are in fact the results of performing the equivalent *direct* operations $\frac{(1+\Delta)^x - 1}{\Delta}$ and $\frac{1-(1+\Delta)^{-x}}{\Delta}E^x$ on v_0.

Here all the operations performed on v_a are direct, and the result is given in differences of the first term.

Ex. 3. To find the sum of x terms of the series $1^2+2^2+\ldots$

Applying* (7) we have (since $\Delta v_0=1,\ \Delta^2 v_0=2$)

$$1^2+2^2+\ldots+(x-1)^2=\Sigma x^2=C+\frac{x(x-1)}{2}+\frac{x(x-1)(x-2)}{1.2.3}.2.$$

Putting $x=2$ we see that C is zero, and adding x^2 to both sides we obtain

$$1^2+2^2+\ldots+x^2=x^2+\frac{x(x-1)}{2}+\frac{x(x-1)(x-2)}{3}$$

$$=\frac{x(x+1)(2x+1)}{6}.$$

Ex. 4. Find the sum of n terms of the series whose n^{th} term is n^3+7n.

We shall here apply formula (9).

The first terms are	8	22	48	92	......
,, ,, differences ,,		14	26	44	....
,, second ,, ,,			12	18	....
,, third ,, ,,				6	....

$$\therefore \text{sum of } n \text{ terms}=8n+14\frac{n(n-1)}{1.2}$$

$$+12\frac{n(n-1)(n-2)}{1.2.3}+6\frac{n(n-1)(n-2)(n-3)}{1.2.3.4}.$$

4th Form. Any rational fraction of the form

$$\frac{\phi(x)}{u_x u_{x+1}\ldots u_{x+m}},$$

* In practice it will be found better to resolve the nth term into factorials and apply the rule given in the note to page 68.

u_x being of the form $ax+b$, and $\phi(x)$ a rational and integral function of x of a degree lower by at least two unities than the degree of the denominator.

Expressing $\phi(x)$ in the form

$$\phi(x) = A + Bu_x + Cu_x u_{x+1} + \ldots + Eu_x u_{x+1} \ldots u_{x+m-2},$$

$A, B \ldots$ being constants to be determined by equating coefficients, or by an obvious extension of the theorem of Chap. II. Art. 5, we find

$$\Sigma \frac{\phi(x)}{u_x u_{x+1} \ldots u_{x+m}} = A\Sigma \frac{1}{u_x u_{x+1} u_{x+2} \ldots u_{x+m}} + B\Sigma \frac{1}{u_{x+1} u_{x+2} u_{x+3} \ldots u_{x+m}}$$

$$+ \ldots + E\Sigma \frac{1}{u_{x+m-1} u_{x+m}},$$

and each term can now be integrated by (5).

Again, supposing the numerator of a rational fraction to be of a degree less by at least two unities than the denominator, but intermediate factors alone to be wanting in the latter to give to it the factorial character above described, then, these factors being supplied to both numerator and denominator, the fraction may be integrated as in the last case.

Ex. 5. Thus u_x still representing $ax+b$, we should have

$$\Sigma \frac{x}{u_x u_{x+2} u_{x+3}} = \Sigma \frac{x u_{x+1}}{u_x u_{x+1} u_{x+2} u_{x+3}},$$

with the second member of which we must proceed as before.

Ex. 6. Find the sum of n terms of the series

$$\frac{2}{1.3.4} + \frac{3}{2.4.5} + \ldots.$$

Here the nth term

$$= \frac{n+1}{n(n+2)(n+3)} = \frac{n^2+2n+1}{n(n+1)(n+2)(n+3)}$$

$$= \frac{n(n+1)+n+1}{n(n+1)(n+2)(n+3)} = \frac{1}{(n+2)(n+3)}$$

$$+ \frac{1}{(n+1)(n+2)(n+3)} + \frac{1}{n(n+1)(n+2)(n+3)}.$$

The sum of n terms therefore, by the rule on page 68,

$$= C - \frac{1}{n+3} - \frac{1}{2(n+2)(n+3)} - \frac{1}{3(n+1)(n+2)(n+3)}$$

$$= C - \frac{6n^2+21n+17}{6(n+1)(n+2)(n+3)},$$

and C can easily be shewn to equal $\frac{17}{36}$.

We thus can find the sum of n terms of any series whose n^{th} term is $\phi(n)$, provided that $\phi(n)$ be either (1) a rational integral function of n, or (2) a fraction whose denominator is the product of terms of an arithmetical series that remain a constant distance from the nth term, and whose numerator is of a degree lower by at least two than its denominator*.

5th Form. Functions of the form a^x or $a^x\phi(x)$ where $\phi(x)$ is rational and integral.

* Since $\phi(n)\,\epsilon^{nx} = \phi(D)\,\epsilon^{nx}$ we may write

$$\phi(a) + \phi(a+1) + \ldots \phi(a+n-1) = [\phi(D)\,\{\epsilon^{ax} + \epsilon^{(a+1)x} + \ldots \epsilon^{(a+n-1)x}\}]_{x=0}$$

$$= \left[\phi(D)\left\{\frac{\epsilon^{(a+n)x} - \epsilon^{ax}}{\epsilon^x - 1}\right\}\right]_{x=0},$$

and the series may therefore be summed by the methods of Differential Calculus or Differential Equations according as $\phi(n)$ is an integral function of n or not. That the result thus obtained is identical with that in the text follows from the identity demonstrated in (16) page 23, viz.

$$\phi(D)\,\psi(0) = \psi(D)\,\phi(0).$$

For this gives

$$\phi(D)\left\{\frac{\epsilon^{(a+n)0} - \epsilon^{a0}}{\epsilon^0 - 1}\right\} = \frac{\epsilon^{(a+n)D} - \epsilon^{aD}}{\epsilon^D - 1}\,\phi(0) = \frac{E^{a+n} - E^a}{E-1}\,\phi(0)$$

$$= \Sigma\,\{\phi(a+n) - \phi(a)\} = \Sigma\phi(a+n) - \Sigma\phi(a),$$

which agrees with the previous expression.

From (13) page 8, we obtain at once $\Sigma a^x = \dfrac{a^x}{a-1}$. For the integration of $a^x\phi(x)$ we shall have recourse to symbolical methods.

$$\begin{aligned}
\Sigma a^x\phi(x) &= \Delta^{-1}a^x\phi(x)\\
&= (\epsilon^D - 1)^{-1}\,\epsilon^{x\log a}\phi(x)\\
&= \epsilon^{x\log a}\,(\epsilon^{D+\log a} - 1)^{-1}\phi(x)^*\\
&= a^x\,(a\epsilon^D - 1)^{-1}\phi(x) = a^x\,\{a\,(1+\Delta) - 1\}^{-1}\phi(x)\\
&= \frac{a^x}{a-1}\left(1 + \frac{a\Delta}{a-1}\right)^{-1}\phi(x)\\
&= \frac{a^x}{a-1}\left\{\phi(x) - \frac{a}{a-1}\Delta\phi(x)\right.\\
&\qquad \left. + \frac{a^2}{(a-1)^2}\Delta^2\phi(x) - \ldots\right\} \qquad (10),
\end{aligned}$$

to which of course an arbitrary constant must be added.

It will be found that the direct application of this theorem† is the simplest method of summing such series as have their x^{th} term of the form $a^x.\phi(x)$.

* By means of the well-known formula $f(D)\epsilon^{mx}\,\phi(x) = \epsilon^{mx}f(D+m)\,\phi(x)$. The proof of this formula is given in Boole's *Diff. Eq.* (First Ed., p. 385), and in many other books.

† The demonstration of (10) can be still farther simplified by quoting the theorem,

$$f(E)\,a^x\,\phi(x) = a^x f(aE)\,\phi(x).$$

This may be deduced from the formula above quoted, but is more readily demonstrated independently, since if $A_n E^n$ be one term of the expansion of $f(E)$ in powers of E we have

$$A_n\,E^n\,a^x\phi(x) = A_n a^{x+n}\,\phi(x+n) = a^x\,.\,A_n\,a^n\,E^n\,\phi(x) = a^x\,.\,A_n\,(aE)^n\,\phi(x),$$

summing all such terms we get

$$f(E)\,a^x\phi(x) = a^x f(aE)\,\phi(x),$$

and the demonstration of (10) runs thus,

$$\begin{aligned}
\Delta^{-1}\,a^x\phi(x) &= (E-1)^{-1}\,a^x\phi(x) = a^x(aE-1)^{-1}\,\phi(x)\\
&= a^x\,\{a\,(1+\Delta) - 1\}\,\phi(x) = \ldots.
\end{aligned}$$

Ex. 7. Find the sum of the series

$$1^2 . 2^1 + 2^2 . 2^2 + 3^2 . 2^3 + \dots.$$

Sum to n terms

$$= n^2 . 2^n + \Sigma n^2 . 2^n$$

$$= n^2 . 2^n + \frac{2^n}{2-1}\left\{n^2 - \frac{2}{2-1}\Delta n^2 + \frac{4}{(2-1)^2}\Delta^2 n^2\right\} + C$$

$$= 2^n \{2n^2 - 4n + 6\} + C.$$

The method just given may be generalized to apply to all functions of the form $u_x . \phi(x)$, where $\phi(x)$ is rational and integral, and u_x is a function such that we know the value of $\Delta^{-n} u_x$ for all integral values of n. In this case we have (comp. Ex. 3, p. 20)

$$\Sigma u_x \phi(x) = (EE' - 1)^{-1} u_x \phi(x) = (\Delta E' + \Delta')^{-1} u_x \phi(x)$$

(E' and Δ' being supposed to operate on ϕ and E and Δ on u_x alone)

$$= \frac{1}{\Delta E'}\left\{1 - \frac{\Delta'}{\Delta E'} + \frac{\Delta'^2}{\Delta^2 E'^2} - \dots\right\} u_x \phi(x)$$

$$= \Delta^{-1} u_x . \phi(x-1) - \Delta^{-2} u_x . \Delta\phi(x-2)$$
$$+ \Delta^{-3} u_x . \Delta^2 \phi(x-3) - \dots \qquad (11),$$

dropping the accents as no longer necessary.

Ex. 8. A good example of the use of the above formula is got by taking $u_x \equiv \sin(ax+b)$. From (17), page 8, we get easily

$$\Delta^{-n} \sin(ax+b) = \frac{\sin\left\{ax + b - \dfrac{n(a+\pi)}{2}\right\}}{\left(2 \sin \dfrac{a}{2}\right)^n}.$$

Let us take then the series whose n^{th} term is

$$(n-7)\sin(an+b);$$

the sum of n terms will be

$$(n-7)\sin(an+b)+\Sigma(n-7)\sin(an+b)$$

$$=(n-7)\sin(an+b)+(n-8)\,.\frac{\sin\left(an+b-\frac{a+\pi}{2}\right)}{2\sin\frac{a}{2}}$$

$$-\frac{\sin\{an+b-(a+\pi)\}}{\left(2\sin\frac{a}{2}\right)^2}+C.$$

6th. *Miscellaneous Forms.* When a function proposed for integration cannot be referred to any of the preceding forms, it will be proper to divine if possible the *form* of its integral from general knowledge of the effect of the operation Δ, and to determine the constants by comparing the difference of the conjectured integral with the function proposed.

Thus since

$$\Delta a^x\phi(x)=a^x\psi(x),$$

where $\psi(x)=a\phi(x+1)-\phi(x)$, it is evident that if $\phi(x)$ be a rational fraction $\psi(x)$ will also be such. Hence if we had to integrate a function of the form $a^x\psi(x)$, $\psi(x)$ being a rational fraction, it would be proper to try first the hypothesis that the integral was of the form $a^x\phi(x)$, $\phi(x)$ being a rational fraction the constitution of which would be suggested by that of $\psi(x)$.

Thus also, since $\Delta\sin^{-1}\phi(x)$, $\Delta\tan^{-1}\phi(x)$, &c., are of the respective forms $\sin^{-1}\psi(x)$, $\tan^{-1}\psi(x)$, &c., $\psi(x)$ being an algebraic function when $\phi(x)$ is such, and, in the case of $\tan^{-1}\phi(x)$, rational if $\phi(x)$ be so, it is usually not difficult to conjecture what must be the forms, if finite forms exist, of

$$\Sigma\sin^{-1}\psi(x),\quad \Sigma\tan^{-1}\psi(x),\ \ldots,$$

$\psi(x)$ being still supposed algebraic.

The above observations may be generalized. The operation denoted by Δ does not change or annul the *functional* characteristics of the subject to which it is applied. It does not convert transcendental into algebraic functions, or one species of transcendental functions into another. And thus, in the inverse procedure of integration, the limits of conjecture are narrowed. In the above respect the operation Δ is unlike that of differentiation, which involves essentially a procedure to the limit, and in the limit new forms arise.

Instances of the above will be given in the Examples at the end of the chapter, but we subjoin the following by way of illustration.

Ex. 9. To sum, when possible, the series

$$\frac{1^2 . x}{2 . 3} + \frac{2^2 . x^2}{3 . 4} + \frac{3^2 . x^3}{4 . 5} + \ldots \text{ to } n \text{ terms.}$$

The nth term, represented by u_n, being $\dfrac{n^2 . x^n}{(n+1)(n+2)}$, we have

$$\Sigma u_{n+1} = \frac{n^2 x^n}{(n+1)(n+2)} + \Sigma \frac{n^2 x^n}{(n+1)(n+2)}.$$

Now remembering that the summation has reference to n, assume

$$\Sigma \frac{n^2 x^n}{(n+1)(n+2)} = \frac{an+b}{n+1} x^n.$$

Then, taking the difference, we have

$$\frac{x^n n^2}{(n+1)(n+2)} = x^n \left\{ x \frac{a(n+1)+b}{n+2} - \frac{an+b}{n+1} \right\}$$

$$= x^n \frac{a(x-1)n^2 + (2a+b)(x-1)n + (a+b)x - 2b}{(n+1)(n+2)}.$$

That these expressions may agree we must have

$$a(x-1) = 1, \quad (2a+b)(x-1) = 0, \quad (a+b)x - 2b = 0.$$

Whence we find

$$x = 4, \quad a = \frac{1}{3}, \quad b = -\frac{2}{3}.$$

The proposed series is therefore integrable if $x=4$*, and we have

$$\Sigma \frac{4^n \cdot n^2}{(n+1)(n+2)} = \frac{1}{3} \cdot \frac{n-2}{n+1} \cdot 4^n + C.$$

Substituting, determining the constant, and reducing, there results

$$\frac{1^2 \cdot 4}{2 \cdot 3} + \frac{2^2 \cdot 4^2}{3 \cdot 4} \cdots + \frac{n^2 4^n}{(n+1)(n+2)} = \frac{4^{n+1}}{3} \cdot \frac{n-1}{n+2} + \frac{2}{3}.$$

3. Σ is of course, like Δ, E, and D, an operation capable of repetition and therefore obeying the index-law; $\Sigma^2 u_x$ being defined as $\Sigma(\Sigma u_x)$. Our symbolical methods will render it an easy matter to obtain expressions for Σ^n (or Δ^{-n}) analogous to those already obtained for Σ, but we shall have to add, as in Integral Calculus, a function of the form

$$C_0 + C_1 x + \ldots + C_{n-1} x^{n-1}$$

(where C_0, C_1, ... are arbitrary or undetermined constants) instead of the single arbitrary constant which we added in the previous instance. We shall merely give the formula for Σ^n analogous to (10) and leave the others as an exercise for the ingenuity of the student. It is

$$\Sigma^n a^x \phi(x) = \frac{a^x}{(a-1)^n} \left\{ \phi(x) - n \frac{a}{a-1} \Delta\phi(x) + \frac{n(n+1)}{1.2} \cdot \left(\frac{a}{a-1}\right)^2 \Delta^2 \phi(x) - \ldots \right\} + C_0 + C_1 x + \ldots + C_{n-1} x^{n-1} \quad (12).$$

* The explanation of this peculiarity is very easy:

$$u_n \equiv \frac{n^2 x^n}{(n+1)(n+2)} = \left\{1 - \frac{4}{n+2} + \frac{1}{n+1}\right\} x^n,$$

and the summation of the above series would require a finite expression for $\Sigma \frac{x^n}{n}$ if x had not such a value that the term $\frac{x^{r+1}}{r+2}$ which occurs in the $(r+1)^{\text{th}}$ term exactly cancelled the term $\frac{-4x^r}{r+2}$ that occurs in the rth term, i.e. unless $x=4$.

It will be found that the 1st, 3rd, and 5th forms can have their n^{th} Finite Integrals expressed in finite terms, but that the 2nd and 4th only permit of this if n be not too great.

Conditions of extension of direct to inverse forms. Nature of the arbitrary constants.

4. From the symbolical expression of Σ in the forms $(\epsilon^D - 1^{-1})$, and more generally of Σ^n in the form $(\epsilon^D - 1)^{-n}$, flow certain theorems which may be regarded as extensions of some of the results of Chap. II. To comprehend the true nature of these extensions the peculiar *interrogative* character of the expression $(\epsilon^{\frac{d}{dx}} - 1)^{-n} u_x$ must be borne in mind. Any legitimate transformation of this expression by the development of the symbolical factor must be considered, in so far as it consists of direct forms, to be an *answer* to the question which that expression proposes; in so far as it consists of inverse forms to be a replacing of that question by others. But the answers will not be of necessity sufficiently general, and the substituted questions if answered in a perfectly unrestricted manner may lead to results which are too general. In the one case we must introduce arbitrary constants, in the other case we must determine the connecting relations among arbitrary constants; in both cases falling back upon our prior knowledge of what the character of the true solution must be. Two examples will suffice for illustration.

Ex. 1. Let us endeavour to deduce symbolically the expression for Σu_x, given in (3), Art. 1.

Now
$$\begin{aligned}\Sigma u_x &= (E-1)^{-1} u_x \\ &= (E^{-1} + E^{-2} + \&c.)\, u_x \\ &= u_{x-1} + u_{x-2} + u_{x-3} \dots + \dots.\end{aligned}$$

Now this is only a particular form of Σu_x corresponding to $a = -\infty$ in (3). To deduce the general form we must add an arbitrary constant, and if to that constant we assign the value

$$-(u_{a-1} + u_{a-2} \dots + \dots),$$

we obtain the result in question.

Ex. 2. Let it be required to develope $\Sigma u_x v_x$ in a series proceeding according to Σv_x, $\Sigma^2 v_x$, ...

We have by (11), page 74,

$$\Sigma u_x v_x = u_{x-1}\,\Sigma v_x - \Delta u_{x-2}\,\Sigma^2 v_x + \Delta^2 u_{x-3}\,\Sigma^3 v_x - \ldots.$$

In applying this theorem, we are not permitted to introduce unconnected arbitrary constants into its successive terms. If we perform on both sides the operation Δ, we shall find that the equation will be identically satisfied provided $\Delta\Sigma^n u_x$ in any term is equal to $\Sigma^{n-1} u_x$ in the preceding term, and this imposes the condition that the constants in $\Sigma^{n-1} u_x$ be retained without change in $\Sigma^n u_x$. And as, if this be done, the equation will be satisfied, it follows that however many those constants may be, they will *effectively* be reduced to one. Hence then we may infer that if we express the theorem in the form

$$\Sigma u_x v_x = C + u_{x-1}\,\Sigma v_x - \Delta u_{x-1}\,\Sigma^2 v_x + \Delta^2 u_{x-2}\,\Sigma^3 v_x \quad (1),$$

we shall be permitted to neglect the constants of integration, provided that we always deduce $\Sigma^n v_x$ by direct integration from the value of $\Sigma^{n-1} v_x$ in the preceding term.

If u_x be rational and integral, the series will be finite, and the constant C will be the one which is due to the last integration effected.

We have seen that C is a constant as far as Δ is concerned, i.e. that $\Delta C = 0$. It is therefore a periodical constant going through all its values during the time that x takes to increase by unity. The necessity of a periodical constant C to complete the value of Σu_x may also be established, and its analytical expression determined, by transforming the problem of summation into that of the solution of a differential equation.

Let $\Sigma u_x = y$, then y is solely conditioned by the equation $\Delta y = u_x$, or, putting $\epsilon^{\frac{d}{dx}} - 1$ for Δ, by the linear differential equation

$$(\epsilon^{\frac{d}{dx}} - 1)\, y = u_x.$$

Now, by the theory of linear differential equations, the complete value of y will be obtained by adding to any particular value v_x the complete value of what y would be, were u_x equal to 0. Hence

$$\Sigma u_x = v_x + C_1\epsilon^{m_1x} + C_2\epsilon^{m_2x} + \ldots \quad (2),$$

$C_1, C_2, \ldots$ being arbitrary constants, and $m_1, m_2, \ldots$ the different roots of the equation

$$\epsilon^m - 1 = 0.$$

Now all these roots are included in the form

$$m = \pm 2i\pi\sqrt{-1},$$

i being 0 or a positive integer. When $i=0$ we have $m=0$, and the corresponding term in (2) reduces to a constant. But when i is a positive integer, we have in the second member of (2) a pair of terms of the form

$$C\epsilon^{2i\pi\sqrt{-1}} + C'\epsilon^{-2i\pi\sqrt{-1}},$$

which, on making $C + C' = A_i$, $(C - C')\sqrt{-1} = B_i$, is reducible to $A_i \cos 2i\pi + B_i \sin 2i\pi$. Hence, giving to i all possible integral values,

$$\Sigma u_x = v_x + C + A_1 \cos 2\pi x + A_2 \cos 4\pi x + A_3 \cos 6\pi x + \ldots$$
$$+ B_1 \sin 2\pi x + B_2 \sin 4\pi x + B_3 \sin 6\pi x + \ldots \quad (3).$$

The portion of the right-hand member of this equation which follows v_x is the general analytical expression of a periodical constant as above defined, viz. as ever resuming the same value for values of x, whether integral or fractional, which differ by unity. It must be observed that when we have to do, as indeed usually happens, with only a particular set of values of x progressing by unity, and not with all possible sets, the periodical constant merges into an ordinary, i.e. into an absolute constant. Thus, if x be exclusively integral, (3) becomes

$$\Sigma u_x = v_x + C + A_1 + A_2 + A_3 + \ldots$$
$$= v_x + c,$$

c being an absolute constant.

It is usual to express periodical constants of equations of differences in the form $\phi(\cos 2\pi x, \sin 2\pi x)$. But this notation is not only inaccurate, but very likely to mislead. It seems better either to employ C, leaving the interpretation to the general knowledge of the student, or to adopt the correct form

$$C + \Sigma_i (A_i \cos 2i\pi x + B_i \sin 2i\pi x) \qquad (4).$$

We shall usually do the former.

5. The student will doubtless already have perceived how much the branch of mathematics that forms the subject of our present consideration suffers from its not possessing a clear and independent set of technical terms. It is true that by its borrowing terms from the Infinitesimal Calculus to supply this want, we are continually reminded of the strong analogies that exist between the two, but in scientific language accuracy is of more value than suggestiveness, and the closeness of the affinity of the analogous processes is by no means such that it is profitable to denote them by the same terms. The shortcomings of the nomenclature of the subject will be felt at once if one thinks of the phrases which describe the operations analogous to the three chief operations in the Infinitesimal Calculus, i.e. Differentiation, Integration, and Integration between limits. There is no reason why the present state of confusion should be permanent, so that we shall in future (in the notes at least) denote these by the unambiguous phrases, *performing* Δ, *taking the Difference-Integral* (or performing Σ), and *summing*, and shall name the two divisions of the calculus, the *Difference-* and the *Sum-Calculus* respectively, and consider them as together forming the *Finite Calculus*. The preceding chapters have been occupied with the Difference-Calculus exclusively—the present is the first in which we have approached problems analogous to those of the Integral Calculus; for it must be borne in mind that such problems as those on Quadratures are merely instances of use being made of the results of the Difference-Calculus, and have nothing to do with the Sum-Calculus, except perhaps in the case of the formula on page 55. Enough has been said about the analogy of the various parts of our earlier chapters with corresponding portions of the Differential Calculus, and we shall here speak only of the exact nature and relations of the Sum-Calculus.

If the nth term of a series be known, and its sum be required, it is tantamount to seeking the difference-integral, and our power of finding the difference-integral is coextensive with our power of finding the sum of any number of terms. Hence the summation of all series, whose sum to n terms can be obtained, is the work of the Sum-Calculus. It is true that there are many series, that can be summed by an artifice, of which we have taken no notice, but that is not because they do not belong to our subject, but because they are too isolated to be important. But it must be remembered that the difference-integral is only obtainable when we can find the sum of any number of consecutive terms we may wish.

But there are many cases in which we seek the sum of n terms of a series which is such that *each term of the series involves* n, e.g. we might desire the sum of the series $1 \,.\, n + 2 \,.\, (n-1) + 3 \,.\, (n-2) + \ldots$ to n terms. Now in a certain sense this is not a case of summation; we do not seek the sum of *any* number of terms, but of a *particular* number of terms depending on the first term of the series itself. And, as might be expected, this operation has not the close connexion that we previously found with that of finding the difference-integral of any term; for though the knowledge of the latter would enable us to sum the series, yet the knowledge of the sum of the series will not enable us to find the difference-integral of any term. These must be called *definite* difference-integrals, and hold exactly the same position that Definite Integrals occupy in the Infinitesimal Calculus. No one would think of excluding from the domain of Integral Calculus the treatment of such functions as the definite integral $\int_0^a x^l (a-x)^m dx$, because the knowledge of its value does not give us any clue to that of the indefinite integral $\int x^l (a-x)^m dx$, and is obtained indirectly without its being made to depend on our first arriving at the knowledge of the latter.

By similar considerations we shall arrive at a right view of the relation of infinite series to the Sum-Calculus. It is often supposed that it has nothing to do with such series—that the summation of finite series is its business, and that this is wholly distinct from the summation of infinite series. This is by no means correct. The true statement is that such series are *definite* difference-integrals, whose upper limit is ∞, and so far they as much belong to our subject as $\int_0^\infty \epsilon - x^2 dx$ does to the Infinitesimal Calculus.

How is it then that the whole subject of series is not referred to this Calculus, but is separated into innumerable portions, and treated of in all imaginable connexions? It is that in *the expression of such series* as those we are speaking of, reference being only made to *finite* quantities, there *is nothing to distinguish them from ordinary algebraical expressions*, except that the symmetry is so great that only a few terms need be written down. Hence when it is summed by an artifice, and not by direct use of the laws of the Sum-Calculus, there is nothing to distinguish the process from an ordinary algebraical transformation or demonstration of the identity of two different expressions. Now in Definite Integrals that are similarly evaluated by an artifice, there is perhaps just as little claim for the evaluation to be classed as a process belonging to the Infinitesimal Calculus, but the *expression of the subject of that process involving the notation and fundamental ideas of the Calculus*, it is naturally classed along with processes that really belong to the Calculus. Thus the Infinitesimal Calculus has a wide field to which no recognized branch of the Finite Calculus corresponds, not because it does not exist, but because it is not reserved for treatment here. No doubt this has its disadvantages. Series would be more systematically treated, and the processes of summation more fully generalized, if they were dealt with collectively; yet on the other hand it is a great advantage in the Finite Calculus to have to do only with such processes as really depend on its laws, and not with processes that are really foreign to it, and are only connected therewith by the fact that their subject-matter in these particular instances is expressed in the form of a series, i.e. in the notation of the Calculus.

It is not usual to speak of such identities as Definite Difference-Integrals, but a certain class of them are considered in this light in a paper by Libri (*Crelle*, XII. 240).

Before leaving the subject of Definite Difference-Integrals we must mention a paper by Leslie Ellis (*Liouville*, IX. 422), in which he demonstrates a theorem analogous to the well-known one on the value of

$$\iiiint \ldots f(x+y+\ldots)\, dx\, dy\, dz \ldots,$$

where $x+y+z+\ldots \leqq 1$. The method is a very beautiful one, but we must not be supposed to endorse it as rigorous, since one part involves the evaluation of $\sum\limits_0^\infty x^{(p)} \cos ax$.

The fundamental operations of the Finite Calculus are taken as Δ with its correlative Σ. In this view of the subject the sign of each term is supposed to be $+$, not that its algebraical value is supposed to be positive, but that its sign must be accounted for by its form. Thus if we take the series $u_0 - u_1 + u_2 - \ldots$, we must call the general term $(-1)^x u_x$. To avoid this complication in the treatment of series whose terms are alternately positive and negative, some have wished to have a second Calculus whose fundamental operation is $\zeta \equiv 1 + E$, the correlative of which, ζ^{-1}, would of course denote the operation of summing such a series. A series of papers by Oettinger, the inventor of it, will be found in *Crelle*, Vols. XI.—XVI. In these he developes the new Calculus in a manner strictly analogous to that in which he subsequently treats the Difference-Calculus, connects them similarly with the Infinitesimal Calculus, demonstrates analogous formulæ, and applies them at first to simple cases and then to more complex ones, especially to those series whose terms are products of the more simple functions and those most suitable to such treatment. The work is unsymbolical, and therefore clumsy and tedious compared with more recent work, and we should not have referred to the papers here (for we consider it highly unadvisable to invent a new Calculus for a comparatively unimportant class of questions that can very easily be dealt with by our present methods) were it not that his results are very copious and detailed. The student who desires practice in the symbolical methods cannot do better than take one of these papers and employ himself in demonstrating by such methods the results there given. Should he desire however a statement of the nature and advantages of this more elaborate treatment of series, he will find it in a review by Oettinger. (Grunert, *Archiv*. XIII. 36.)

This is not the only attempt to introduce a new Finite-Calculus. A certain class of series is treated in a paper by Werner (Grunert, *Archiv*. XXII. 264), by means of a calculus whose fundamental operation, $\Lambda \equiv E - v_x$, is almost the most general form of linear fundamental operation that can be imagined.

EXERCISES.

1. Sum to n terms the following series:

$$1 . 3 . 5 . 7 + 3 . 5 . 7 . 9 + \ldots$$

$$\frac{1}{1 . 3 . 5 . 7} + \frac{1}{3 . 5 . 7 . 9} + \ldots$$

$$1.3.5.10+3.5.7.12+5.7.9.14+\ldots$$

$$\frac{10}{1.3.5}+\frac{12}{3.5.7}+\frac{14}{5.7.9}+\ldots$$

$$1.3.5.\cos\theta+3.5.7.\cos 2\theta+5.7.9.\cos 3\theta+\ldots$$

$$1+2a\cos\theta+3a^2\cos 2\theta+4a^3\cos 3\theta+\ldots$$

2. The successive orders of figurate numbers are defined by this;—that the xth term of any order is equal to the sum of the first x terms of the order next preceding, while the terms of the first order are each equal to unity. Shew that the xth term of the nth order is

$$\frac{x(x+1)\ \ldots\ (x+n-2)}{\lfloor n-1}.$$

3. If $\Sigma'_n u_x$ denote the sum of the first n terms of the series u_0, u_2, u_4, &c. shew that

$$\Sigma'_n u_x = \frac{1}{2}\left\{\Sigma - \frac{1}{2} + \frac{\Delta}{4} - \frac{\Delta^2}{8} + \ldots\right\}(u_{2n} - u_0),$$

and apply this to find the sum of the series

$$1.3.5+5.7.9+9.11.13+\ldots.$$

4. Expand $\Sigma\phi(x)\cos mx$ in a series of differences of $\phi(x)$.

5. Find in what cases, when u_x is one of the five forms given as integrable in the present Chapter, we can find the sum of n terms of the series

$$u_0 - u_1 + u_2 - u_3 + \ldots,$$

and construct the suitable formulæ in each case.

6. Sum the following series to n terms:

$$\frac{1}{\sin\theta}+\frac{1}{\sin 2\theta}+\frac{1}{\sin 4\theta}+\ldots$$

$$\frac{1}{\cos\theta.\cos 2\theta}+\frac{1}{\cos 2\theta.\cos 3\theta}+\ldots$$

7. Shew that $\cot^{-1}(p+qn+rn^2)$ is integrable in finite terms whenever

$$q^2-r^2=4.(pr-1).$$

Obtain

$$\Sigma\tan^{-1}\frac{x}{1+n(n-1)x^2},\ \text{and}\ \Sigma\frac{\log\tan 2^n\theta}{2^n},\ \text{and}\ \Sigma\frac{2^n(n-1)}{n(n+1)}.$$

8. It is always possible to assign such values to s, real or imaginary, that the function

$$\frac{(\alpha+\beta x+\gamma x^2+\ldots+\nu x^n)\,s^x}{u_x u_{x+1}\ldots u_{x+m-1}}$$

shall be integrable in finite terms; $\alpha, \beta, \ldots \nu$ being any constants and $u_x \equiv ax+b$.

(Herschel's *Examples of Finite Differences*, p. 47.)

9. Shew that

$$u_0+u_1\cos 2\theta+u_2\cos 4\theta+\ldots=\frac{u_0}{2}-\frac{\Delta u_0}{4\sin^2\theta}$$

$$+\frac{\Delta^2 u_0}{8\sin^3\theta}.\sin\theta+\frac{\Delta^3 u_0}{16\sin^4\theta}\cos 2\theta-\frac{\Delta^4 u_0}{32\sin^5\theta}\sin 3\theta-\ \ldots.$$

10. If $\Delta u_x \equiv u_{x+h}-u_x$ and $\lambda=\dfrac{a^h}{a^h-1}$, shew that

$$u_x+\lambda\Delta u_x+\lambda^2\Delta^2 u_x+\ \ldots+\lambda^n\Delta^n u_x$$

$$=a^{-x}\left\{(a^h-1)\,\Sigma a^x u_x+\lambda^n\Sigma a^{x+h}\Delta^{n+1}u_x\right\}.$$

Find the sum of n terms of the series whose n^{th} terms are

$$(a+n-1)^m x^{n-1}\ \text{and}\ (a+n-1)^{(m)}x^{n-1}.$$

11. Prove the theorem

$$\Sigma^n u_x v_x=u_x\Sigma^n v_x-n\Delta u_x\Sigma^{n+1}v_{x+1}+\frac{n(n+1)}{1.2}\Delta^2 u_x\Sigma^{n+2}v_{x+2}-\ \ldots.$$

12. If $\phi(x)\equiv v_0+v_1x+v_2x^2+\ \ldots$, shew that

$$u_0v_0+u_1v_1x+u_2v_2x^2+\&c.=u_0\phi(x)+x\phi'(x).\Delta u_0$$

$$+\frac{x^2}{1.2}\phi''(x).\Delta^2u_0+\ \ldots;$$

and if $\phi(x) \equiv v_0 + v_1 x + v_2 x^{(2)} + \dots$, then

$$u_0 v_0 + u_1 v_1 x + u_2 v_2 x^{(2)} + \dots = u_0 \phi(x) + x \Delta \phi(x) \, . \, \Delta v_0 + \frac{x^{(2)}}{1 \, . \, 2} \Delta^2 \phi(x) \, . \, \Delta^2 u_0 + \dots .$$

(Guderman, *Crelle*, VII. 306.)

13. Sum to infinity the series

$$0^r + 1^r \, . \, \frac{z}{x} + 2^r \, . \, \frac{z(z-1)}{x(x-1)} + 3^r \, . \, \frac{z(z-1)(z-2)}{x(x-1)(x-2)} + \dots$$

14. If $\phi(x) \equiv v_0 + v_1 x + v_2 x^2 + \dots$, shew that

$$a_r u_r x^r + a_{r+n} u_{r+n} x^{r+n} + a_{r+2n} u_{r+2n} x_{r+2n} + \dots$$
$$= \frac{1}{n} \{ \Sigma [\alpha^{n-r} \phi(\alpha x)] \, u_0 + \Sigma [\alpha^{n-r+1} \phi'(\alpha x)] \Delta u_0 \, . \, x + \dots \},$$

where α is an nth root of unity.

15. If $1^n + 2^n + \dots + m^n \equiv S_n$ and $m(m+1) = p$, shew that $S_n = p^2 f(p)$ or $(2m+1)\, p f(p)$, according as n is odd or even.

(*Nouvelles Annales*, X. 199.)

CHAPTER V.

THE APPROXIMATE SUMMATION OF SERIES.

1. It has been seen that the finite summation of series depends upon our ability to express in finite algebraical terms the result of the operation Σ performed upon the general term of the series. When such finite expression is beyond our powers, theorems of approximation must be employed. And the constitution of the symbol Σ as expressed by the equation

$$\Sigma = (\epsilon^D - 1)^{-1} \dots (1)$$

renders the deduction and the application of such theorems easy.

Speaking generally these theorems are dependent upon the development of the symbol Σ in ascending powers of D.

But another method, also of great use, is one in which we expand in terms of the *successive differences of some important factor of the general term,* i.e. in ascending powers of Δ, where Δ is considered as operating on one factor alone of the general term, and is no longer the inverse of the Σ we are trying to perform*.

* Let us compare these methods of procedure with those adopted in the Integral Calculus. If $\int \phi(x)\, dx$ cannot be obtained in finite terms it is usual either

(1) To expand $\phi(x)$ in a series proceeding by powers of x and to integrate each term separately;

(2) To develope $\int \phi(x)\, dx$ by Bernoulli's Theorem (i.e. by repeated integration by parts) in a series proceeding by *successive differential coefficients of some factor of the general term;* or

As our results are no longer exact it becomes a matter of the greatest importance to determine how far they differ from the exact results, or, in other words, the degree of approximation attained. But this is usually a difficult task, and in order to lessen the difficulty of the subject to the student, we shall separate such investigations from those which first give us the expansions. The order in which we shall treat the subject will therefore be as follows:

I. We shall obtain symbolical expansions for Σ, Σ^2, ... (Chapters V. and VI.)

II. We shall examine the general question of Convergency and Divergency of Series, to ascertain if we may assume the arithmetical equivalence of the results of performing on u_x the operations that we have just found to be symbolically equivalent. (Ch. VII.)

III. Finding that many of our results do not stand the test we shall proceed to find the *exact theorems* corresponding to them, i.e. to find expressions for the remainder after n terms, and thus we shall reestablish the approximateness of these results. (Ch. VIII.)

(3) To develope $\int \phi(x)\, dx$ in a series proceeding by *successive differences of* ϕx by aid of Laplace's formula for Mechanical Quadrature [(27) page 54], which may be written thus:

$$\int \phi(x)\, dx = C + \Sigma\phi(x) - \frac{1}{12}\Delta\phi(x) + \frac{1}{24}\Delta^2\phi(x) - \ldots \qquad (2).$$

We should therefore expect to find in the Sum-Calculus the corresponding methods, viz.:

(1) To expand u_x in a series proceeding by factorials, and to sum each term separately;

(2) To develope Σu_x in a series proceeding by successive differences of some factor of the general term;

(3) To develope Σu_x in a series proceeding by successive differential coefficients of u_x.

Of these (3) and (2) are those mentioned in the text; (1) is not of much use since the cases in which it can be applied are very few, and no theorems of great generality have been found to enable us to obtain the expansion necessary. Besides the resulting series will usually be highly divergent unless the factorials are inverse ones, i.e. have negative indices, so that the results will not be suitable for giving the approximate values we seek. We shall, however, give some account later on of the results that have been obtained by this method.

We shall now commence the first of these divisions.

2. PROP. I. *To develope Σu_x in a series proceeding by the differential coefficients of u_x.*

Since $\Sigma u_x = (\epsilon^{\frac{d}{dx}} - 1)^{-1} u_x$, we must expand $(\epsilon^{\frac{d}{dx}} - 1)^{-1}$ in ascending powers of $\frac{d}{dx}$, and the *form* of the expansion will be determined by that of the function $(\epsilon^t - 1)^{-1}$. For simplicity we will first deduce a few terms of the expansion and examine somewhat its general form, leaving fuller investigations to the next Chapter.

The function $(\epsilon^t - 1)^{-1}$ is not at once suitable for expansion by Maclaurin's Theorem, since it contains a negative power of t; we shall therefore expand $\frac{t}{\epsilon^t - 1}$ either by Maclaurin's Theorem or by actual division and divide the result by t,

$$\frac{t}{\epsilon^t - 1} = \frac{t}{t + \frac{t^2}{1\,.\,2} + \frac{t^3}{1\,.\,2\,.\,3} + \dots}$$

$$= 1 - \frac{t}{2} + \frac{t^2}{12} - \frac{t^4}{720} + \dots \qquad (3).$$

The term $-\frac{t}{2}$ may be shewn to be the only term in the expansion involving an odd power of t. For

$$\frac{t}{\epsilon^t - 1} + \frac{t}{2} = \frac{t}{2} \cdot \frac{\epsilon^t + 1}{\epsilon^t - 1},$$

which does not change when t is changed into $-t$, and therefore can contain, on expansion, even powers of t alone.

From these results we may conclude that the development of $(\epsilon^t - 1)^{-1}$ will assume the form

$$(\epsilon^t - 1)^{-1} = \frac{A_{-1}}{t} + A_0 + A_1 t + A_3 t^3 + A_5 t^5 + \dots \qquad (4).$$

It is however customary to express this development in the somewhat more arbitrary form

$$(\epsilon^t - 1)^{-1} = \frac{1}{t} - \frac{1}{2} + \frac{B_1}{\lfloor 2} t - \frac{B_3}{\lfloor 4} t^3 + \frac{B_5}{\lfloor 6} t^5 - \dots \qquad (5).$$

The quantities B_1, B_3, .. are called Bernoulli's numbers, and will form the subject of the major part of the next Chapter.

Hence we find

$$\Sigma u_x = C + \int u_x dx - \frac{1}{2} u_x + \frac{B_1}{\lfloor 2} \frac{du_x}{dx} - \frac{B_3}{\lfloor 4} \frac{d^3 u_x}{dx^3} + \dots \qquad (6).$$

Or, actually calculating a few of the coefficients,

$$\Sigma u_x = C + \int u_x dx - \frac{1}{2} u_x + \frac{1}{12} \frac{du_x}{dx} - \frac{1}{720} \frac{d^3 u_x}{dx^3}$$
$$+ \frac{1}{30240} \frac{d^5 u_x}{dx^5} \qquad (7).*$$

The following table contains the values of the first ten of Bernoulli's numbers,

$$B_1 = \frac{1}{6},\ B_3 = \frac{1}{30},\ B_5 = \frac{1}{42},\ B_7 = \frac{1}{30},\ B_9 = \frac{5}{66},$$

* Attention has been directed (*Differential Equations*, p. 376) to the *interrogative* character of inverse forms such as

$$(\epsilon^{\frac{d}{dx}} - 1)^{-1} u_x.$$

The object of a theorem of transformation like the above is, strictly speaking, to determine a function of x such that if we perform upon it the corresponding *direct* operation (in the above case this is $\epsilon^{\frac{d}{dx}} - 1$) the result will be u_x. To the inquiry what that function is, a legitimate transformation will necessarily give a correct but not necessarily the most general answer. Thus C in the second member of (6) is, from the mode of its introduction, the constant of ordinary integration; but for the most general expression of Σu_x C ought to be a periodical quantity, subject only to the condition of resuming the same value for values of x differing by unity. In the applications to which we shall proceed the values of x involved will be integral, so that it will suffice to regard C as a simple constant. Still it is important that the true relation of the two members of the equation (6) should be understood.

$$B_{11}=\frac{691}{2730},\ B_{13}=\frac{7}{6},\ B_{15}=\frac{3617}{510},$$

$$B_{17}=\frac{43867}{798},\ B_{19}=\frac{1222277}{2310} \qquad (8).$$

It will be noted that they are ultimately divergent. It will seldom however be necessary to carry the series for Σu_x further than is done in (7), and it will be shewn that the employment of its convergent portion is sufficient.

Applications.

3. The general expression for Σu_x in (7), Art. 2, gives us at once the integral of any rational and entire function of x.

Ex. 1. Thus making $u_x \equiv x^4$, we have

$$\Sigma x^4 = C + \int x^4 dx - \frac{1}{2}x^4 + \frac{1}{12}\frac{d(x^4)}{dx} - \frac{1}{720}\frac{d^3(x^4)}{dx^3}$$

$$= C + \frac{x^5}{5} - \frac{x^4}{2} + \frac{x^3}{3} - \frac{x}{30}.$$

More generally, making $u_x \equiv x^n$ we get

$$\Sigma x^n = C + \frac{x^{n+1}}{n+1} - \frac{1}{2}x^n + \frac{nB_1}{\underline{|2}}x^{n-1} - \frac{n(n-1)(n-2)B_3}{\underline{|4}}x^{n-3}$$

$$- \ldots$$

which at once enables us to connect Bernoulli's numbers with the coefficients of the powers of x in the expression for

$$1^n + 2^n + \ldots\ldots + x^n.$$

But the theorem is of chief importance when finite summation is impossible.

Ex. 2. Thus making $u_x \equiv \frac{1}{x^2}$, we have

$$\Sigma \frac{1}{x^2} = C - \frac{1}{x} - \frac{1}{2x^2} + \frac{1}{12}\left(\frac{-2}{x^3}\right) - \frac{1}{720}\left(-\frac{2.3.4}{x^5}\right) - \dots$$

$$= C - \frac{1}{x} - \frac{1}{2x^2} - \frac{1}{6x^3} + \frac{1}{30x^5} - \dots.$$

The value of C must be determined by the particular conditions of the problem. Thus suppose it required to determine an approximate value of the series

$$\frac{1}{1^2} + \frac{1}{2^2} + \frac{1}{3^2} + \dots \frac{1}{(x-1)^2}.$$

Now by what precedes,

$$\frac{1}{1^2} + \frac{1}{2^2} + \frac{1}{3^2} + \dots \frac{1}{(x-1)^2} = C - \frac{1}{x} - \frac{1}{2x^2} - \frac{1}{6x^3} + \frac{1}{30x^5} - \dots.$$

Let $x = \infty$, then the first member is equal to $\frac{\pi^2}{6}$ by a known theorem, while the second member reduces to C. Hence

$$\frac{1}{1^2} + \frac{1}{2^2} \dots + \frac{1}{(x-1)^2} = \frac{\pi^2}{6} - \frac{1}{x} - \frac{1}{2x^2} - \frac{1}{6x^3} + \frac{1}{30x^5} - \dots$$

and if x be large a few terms of the series in the second member will suffice.

4. When the sum of the series *ad inf.* is unknown, or is known to be infinite, we may approximately determine C by giving to x some value which will enable us to compare the expression for Σu_x, in which the constant is involved, with the actual value of Σu_x obtained from the given series by addition of its terms.

Ex. 3. Let the given series be $1 + \frac{1}{2} + \frac{1}{3} \dots + \frac{1}{x}$.

Representing this series by u_x, we have

$$u_x = \frac{1}{x} + \Sigma \frac{1}{x}$$

$$= \frac{1}{x} + C + \log x - \frac{1}{2x} - \frac{1}{12x^2} + \frac{1}{120x^4} - \dots$$

$$= C + \log x + \frac{1}{2x} - \frac{1}{12x^2} + \frac{1}{120x^4} - \dots.$$

To determine C, assume $x = 10$, then

$$1 + \frac{1}{2} + \frac{1}{3} \dots + \frac{1}{10} = C + \log_\epsilon 10 + \frac{1}{20} - \frac{1}{1200} + \frac{1}{1200000} - \dots.$$

Hence, writing for $\log_\epsilon 10$ its value $2{\cdot}302585$, we have approximately $C = {\cdot}577215$. Therefore

$$u_x = {\cdot}577215 + \log x + \frac{1}{2x} - \frac{1}{12x^2} + \frac{1}{120x^4} - \dots.$$

Ex. 4. Required an approximate value for $1 \,.\, 2 \,.\, 3 \dots x$.

If $u_x \equiv 1 \,.\, 2 \,.\, 3 \dots x$, we have

$$\log u_x = \log 1 + \log 2 + \log 3 \dots + \log x$$

$$= \log x + \Sigma \log x.$$

But $\Sigma \log x = C + \int \log x dx - \frac{1}{2} \log x$

$$+ \frac{B_1}{1 \,.\, 2} \frac{d \log x}{dx} - \frac{B_3}{1 \,.\, 2 \,.\, 3 \,.\, 4} \frac{d^3 \log x}{dx^3} + \dots$$

$$= C + \left(x - \frac{1}{2}\right) \log x - x + \frac{B_1}{1 \,.\, 2x} - \frac{B_3}{3 \,.\, 4x^3} + \frac{B_5}{5 \,.\, 6x^5} - \dots;$$

$$\therefore \log u_x = C + \left(x + \frac{1}{2}\right) \log x - x + \frac{1}{12x} - \dots \qquad (9).$$

To determine C, suppose x very large and tending to become infinite, then

$$\log (1 \,.\, 2 \,.\, 3 \dots x) = C + \left(x + \frac{1}{2}\right) \log x - x,$$

whence

$$1.2.3 \ldots x = \epsilon^{C-x} \times x^{x+\frac{1}{2}} \qquad (10),$$

$$1.2.3 \ldots 2x = \epsilon^{C-2x} \times (2x)^{2x+\frac{1}{2}} \qquad (11).$$

But multiplying (10) by 2^x,

$$2.4.6 \ldots 2x = 2^x \epsilon^{C-x} \times x^{x+\frac{1}{2}} \qquad (12).$$

Therefore, dividing (11) by (12),

$$3.5.7 \ldots (2x-1) = 2^{x+\frac{1}{2}} \epsilon^{-x} x^x,$$

whence

$$\frac{2.4.6 \ldots 2x}{3.5.7 \ldots (2x-1)} = \frac{\epsilon^C x^{\frac{1}{2}}}{2^{\frac{1}{2}}}.$$

But by Wallis's theorem, x being infinite,

$$\frac{2.4.6 \ldots (2x-2)\sqrt{(2x)}}{3.5.7 \ldots (2x-1)} = \sqrt{\left(\frac{\pi}{2}\right)},$$

whence by division

$$\sqrt{(2x)} = \frac{\epsilon^C \sqrt{x}}{\sqrt{\pi}};$$

$$\therefore C = \log \sqrt{(2\pi)}.$$

And now, substituting this value in (9) and determining u_x, we find

$$u_x = \sqrt{(2\pi)} \times x^{x+\frac{1}{2}} \times \epsilon^{-x+\frac{1}{12x}-\frac{1}{360x^3}+\cdots}$$

$$= \sqrt{(2\pi x)} \,.\, x^x \,.\, \epsilon^{-x+\frac{1}{12x}-\frac{1}{360x^3}+\cdots} \qquad (13).$$

If we develope the factor $\epsilon^{\frac{1}{12x}-\frac{1}{360x^3}+\cdots}$ in descending powers of x, we find

$$1.2.3 \ldots x = \sqrt{(2\pi x)} \,.\, x^x \epsilon^{-x} \left(1 + \frac{1}{12x} + \frac{1}{288x^2} - \frac{139}{51840x^3} + \cdots\right) \qquad (14).$$

Hence for very large values of x we may assume

$$1 . 2 . 3 ... x = \sqrt{(2\pi x)} \left(\frac{x}{\epsilon}\right)^x \qquad (15),$$

the ratio of the two members tending to unity as x tends to infinity. And speaking generally it is with the ratios, not the actual values of functions of large numbers, that we are concerned.

Ex. 5. To find an approximate value of $\Gamma(x+1)$ when x is large.

It will be seen that this reduces to the preceding example when x is integral; it has been chosen to illustrate our mode of determining C.

Exactly as in the preceding case we obtain

$$\log u_x = C + \left(x + \frac{1}{2}\right) \log x - x + \frac{B_1}{1 . 2x} - \frac{B_3}{3 . 4x^3} + \frac{B_5}{5 . 6x^5} - \dots \qquad (16),$$

but we can draw no conclusion as to the value of C from the value it bore in (9), nor would any number of special determinations of its value enable us to draw any conclusions as to its general value. But it can be proved (Todhunter's *Int. Cal.* 3rd Ed. p. 254) that

$$\frac{d^2 \log \Gamma x}{dx^2} = \frac{1}{x^2} + \frac{1}{(x+1)^2} + \frac{1}{(x+2)^2} + \dots \textit{ ad. inf.}$$

$$= 0 \text{ when } x \text{ is infinite.}$$

But from (16) we obtain, when x is infinite,

$\dfrac{d^2 \log \Gamma(x+1)}{dx^2} = \dfrac{d^2 C}{dx^2}$, which is therefore zero when x is infinite.

Now C is a *periodic* quantity going through its course of values as x increases by unity—hence $\dfrac{d^2 C}{dx^2}$ is equally periodic.

$$\therefore \frac{d^2 C}{dx^2} = 0,$$

for finite as well as infinite values of x;

$$\therefore C = Ax + B.$$

But C remains unchanged when x is increased by unity; therefore $A = 0$, and C is therefore an absolute constant, and therefore has the value found for it in Ex. 4 when x was an integer, i.e. $C = \log \sqrt{2\pi}$.

Ex. 6. To sum the series

$$1 + \frac{1}{2^{2n}} + \frac{1}{3^{2n}} + \frac{1}{4^{2n}} \dots + \frac{1}{x^{2n}}.$$

Representing the series by u, we have

$$u = \frac{1}{x^{2n}} + \Sigma \frac{1}{x^{2n}}$$

$$= C - \frac{1}{(2n-1)\, x^{2n-1}} + \frac{1}{2x^{2n}} - \frac{2n}{12x^{2n+1}}$$

$$+ \frac{2n\,(2n+1)\,(2n+2)}{720x^{2n+3}} - \dots.$$

For each particular value of n the constant C might be determined approximately as in Ex. 3, but its general expression will be found in Art. 3, Ch. VI.

5. Prop. II. *To develope $\Sigma^n u_x$ in a series proceeding by the differential coefficients of u_x.*

Since $\qquad \Sigma = (\epsilon^D - 1)^{-1}; \quad \therefore \Sigma^n = (\epsilon^D - 1)^{-n},$

and the problem reduces to that of expanding $(\epsilon^t - 1)^{-n}$ in ascending powers of t; or, in other words, to expanding $\frac{t^n}{(\epsilon^t - 1)^n}$ in positive integral powers of t.

Let $$v_n \equiv \frac{1}{(\epsilon^t - 1)^n};$$

$$\therefore \frac{dv_n}{dt} = \frac{-n\epsilon^t}{(\epsilon^t-1)^{n+1}} = \frac{-n(\epsilon^t-1)-n}{(\epsilon^t-1)^{n+1}}$$

$$= -nv_n - nv_{n+1};$$

$$\therefore nv_{n+1} = -\left(\frac{d}{dt}+n\right)v_n.$$

Multiply both sides by $\underline{|n-1}$ and let $w_n \equiv \underline{|n-1}\, v_n$, and the equation becomes

$$w_{n+1} = -\left(\frac{d}{dt}+n\right)w_n = \left(\frac{d}{dt}+n\right)\left(\frac{d}{dt}+n-1\right)w_{n-1}$$
$$= \&c.$$

Ultimately we obtain (writing $n-1$ for n)

$$\underline{|n-1}\,\{\epsilon^t-1\}^{-n} = (-1)^{n-1}\left(\frac{d}{dt}+n-1\right)\left(\frac{d}{dt}+n-2\right)\dots$$
$$\dots\left(\frac{d}{dt}+1\right)\{\epsilon^t-1\}^{-1} \qquad (17).$$

By means of this formula we can obtain developed expressions for Σ^2, Σ^3, ... with great readiness in terms of the coefficients in the expansion of Σ, i.e. in terms of Bernoulli's numbers.

Ex. To develope Σ^3 in terms of D.

From (17),

$$\underline{|2}\,\{\epsilon^t-1\}^{-3} = \left(\frac{d}{dt}+2\right)\left(\frac{d}{dt}+1\right)\{\epsilon^t-1\}^{-1}$$

$$= \left(\frac{d^2}{dt^2}+3\frac{d}{dt}+2\right)\left\{\frac{1}{t}-\frac{1}{2}+A_1 t + A_2 t^2 + \dots\right\} \text{ suppose,}$$

where $A_{2r} = 0$ for all values of r and $A_{2r+1} \equiv (-1)^r \dfrac{B_{2r+1}}{\underline{|2r+2}}$

$$=\frac{2}{t^3}-\frac{3}{t^2}+\frac{2}{t}+(2A_2+3A_1-1)$$

$$+\sum_{r=1}^{r=\infty}[(r+2)(r+1)A_{r+2}+3(r+1)A_{r+1}+2A_r]\,t^r.$$

Hence

$$\Sigma^3 u_x=\iiint u_x\,dx-\frac{3}{2}\iint u_x\,dx+\int u_x\,dx-\frac{3}{8}u_x+\frac{19}{240}\frac{du_x}{dx}-\ \dots.$$

6. Prop. III. *To develope $\Sigma^n u_x$ in a series, proceeding by successive differential coefficients of $u_{x-\frac{n}{2}}$.*

$$\Sigma=\frac{1}{\epsilon^D-1}=E^{-\frac{1}{2}}\frac{\epsilon^{\frac{1}{2}D}}{\epsilon^D-1}=E^{-\frac{1}{2}}\frac{1}{\epsilon^{\frac{1}{2}D}-\epsilon^{-\frac{1}{2}D}}$$

$$\therefore\ D\Sigma=E^{-\frac{1}{2}}\operatorname{cosec}\left(\frac{1}{2}D\sqrt{-1}\right)\times\left(\frac{1}{2}D\sqrt{-1}\right)$$

$$\therefore\ D^n\Sigma^n=E^{-\frac{n}{2}}\operatorname{cosec}^n\left(\frac{1}{2}D\sqrt{-1}\right)\times\left(\frac{1}{2}D\sqrt{-1}\right)^n \qquad (18).$$

Suppose

$$x^n\operatorname{cosec}^n x=1-C_2x^2+C_4x^4-\dots,$$

then

$$\Sigma^n u_x=D^{-n}\left\{1+C_2\left(\frac{D}{2}\right)^2+C_4\left(\frac{D}{2}\right)^4+\dots\right\}u_{x-\frac{n}{2}} \qquad (19)^*.$$

It must be mentioned that the Summation-formula of Art. 2 (which is due to Maclaurin†) is quite as applicable in the form

$$\int u_x\,dx=\Sigma u_x+\frac{1}{2}u_x-\frac{B_1}{\underline{|2}}\frac{du_x}{dx}+\frac{B_3}{\underline{|4}}\frac{d^3u_x}{dx^3}-\dots-C \qquad (20),$$

to the evaluation of integrals by reducing it to a summation, as it is, in its original form, to the summation of series by reducing it to an integration. It is thus a substitute for (27), page 54.

* This remarkably symmetrical expression for Σ^n is due to Spitzer (Grunert, *Archiv.* XXIV. 97).

† *Tract on Fluxions*, 672. Euler gives it also (*Trans. St Petersburg*, 1769), and it is often ascribed to him.

7. PROP. IV. *To expand Σu_x and $\Sigma^n u_x$ in a series proceeding by successive differences of some factor of u_x.*

It will be seen that the formula of (11) page 74 and Ex. 11 page 85, accomplish this object. We shall only treat here of the very important case when $u_x \equiv a^x \phi(x)$ and more especially regard the form which the result takes when $a = -1$, i.e. when the series is

$$\phi(0) - \phi(1) + \phi(2) - \dots .$$

We have in general,

$$\Sigma a^x \phi(x) = (E-1)^{-1} a^x \phi(x) = a^x (aE-1)^{-1} \phi(x) \text{ (note, page 73)}$$

$$= \frac{a^x}{a-1} \left\{1 + \frac{a\Delta}{a-1}\right\}^{-1} \phi(x),$$

which may be now expanded. If $a = -1$, we obtain

$$\Sigma (-1)^x \phi(x) = \frac{(-1)^{x-1}}{2} \left\{1 - \frac{\Delta}{2} + \frac{\Delta^2}{4} - \dots\right\} \phi(x).$$

This enables us to transform many infinite series into others of a more convergent character; for

$$\phi(0) - \phi(1) + \dots \textit{ ad inf.}$$

$$= \frac{1}{1+E} \phi(0) = \frac{1}{2} \left\{1 - \frac{\Delta}{2} + \frac{\Delta^2}{4} - \dots\right\} \phi(0) \qquad (21),$$

which is very rapidly convergent if the other is but slowly so.

Ex. Transform the series $\frac{1}{12} - \frac{1}{13} + \frac{1}{14} - \dots$ into a more convergent form.

Here $\qquad \phi(0) \equiv (0+12)^{(-1)},$

$\therefore$ we have by (21)

$$\frac{1}{12} - \frac{1}{13} + \dots = \frac{1}{2}\left\{\frac{1}{12} + \frac{1}{2.12.13} + \frac{2}{4.12.13.14} + \frac{2.3}{8.12.13.14.15} + \dots\right\},$$

which converges rapidly.

8. It is very often advisable to find the sum of the first few terms of a series by ordinary addition and subtraction, and then to apply our formulæ to the remaining terms, as in this way the convergence of the resulting series is usually greater.

Thus, if we had applied the formula just obtained to the series

$$1-\frac{1}{2}+\frac{1}{3}-\frac{1}{4}+\ldots,$$

we should have obtained

$$\frac{1}{2}\left\{1+\frac{1}{2.1.2}+\frac{2}{4.1.2.3}+\frac{2.3}{8.1.2.3.4}+\ldots\right\},$$

a much more slowly converging series.

This remark is of great importance with reference to all the formulæ of this Chapter. We shall see that the Maclaurin Sum-formula of Art. (2) usually gives rise to series that first converge and then diverge, but that by keeping only the convergent part we obtain an approximate value of the function on the left-hand side of the identity; and also that the closeness of the approximation depends on the smallness of the first of the terms in the rejected portion. From this it follows that by applying the formula in the manner just indicated we can greatly increase the closeness of the approximation. An example will make it clearer.

Ex. Let $u_x \equiv \frac{1}{x^2}$, then the formula becomes

$$\Sigma\frac{1}{x^2}=C-\frac{1}{x}-\frac{1}{2x^2}-\frac{B_1}{x^3}+\frac{B_3}{x^5}-\ \ldots \qquad (22).$$

Taking this between limits ∞ and 1, we obtain

$$1+\frac{1}{4}+\frac{1}{9}+\ldots=1+\frac{1}{2}+B_1-B_3+B_5,\ \ldots.$$

Now, remembering that we must only keep the convergent part of the series, we find that we must stop at B_5, since after that the numbers begin to increase. This gives us 1.65714, the true value being $\frac{\pi^2}{6}$ or 1.64493.

Now let us find the sum thus

$$1+\frac{1}{4}+\frac{1}{9}+\dots \textit{ ad inf.} = 1+\frac{1}{4}+\frac{1}{9}+\frac{1}{16}+\sum_{x=5}^{x=\infty}\frac{1}{x^2}$$

$$=\frac{205}{144}+\frac{1}{5}+\frac{1}{2\,.\,5^2}+\frac{B_1}{5^3}-\frac{B_3}{5^5}+\dots.$$

On examination it will be found that we may in this case keep the terms at least as far as B_{19}*, while the convergence is so rapid at first that by only retaining as far as B_1 we obtain 1.64494. The general advantage of using the formula may be gathered from this example. To obtain an equally close approximation by actual summation, some hundred thousand terms would have to be taken.

9. We can also expand $\Sigma a^x\phi(x)$ in a series proceeding by successive differential coefficients of $\phi(x)$. For

$$\Sigma a^x\phi(x) = (E-1)^{-1}a^x\phi(x) = a^x(aE-1)^{-1}\phi(x) \qquad (23).$$

But by Herschel's Theorem $\psi(e^t) = \psi(E)\,e^{0.t}$,

$\therefore\ \psi(E) = \psi(e^D) = \psi(E')\,e^{0.D}$ as operating factors,

where E' affects 0 only,

$$\therefore\ \Sigma a^x\phi(x) = a^x(aE'-1)^{-1}\left\{1+0\,.\,D+\frac{0^2}{1\,.\,2}D^2+\dots\right\}\phi(x)$$

$$=\frac{a^x}{a-1}\left\{1+A_1\frac{d\phi(x)}{dx}+\frac{A_2}{1\,.\,2}\cdot\frac{d^2\phi(x)}{dx^2}+\dots\right\} \qquad (24),$$

where $A_n \equiv \left\{\frac{aE-1}{a-1}\right\}^{-1}0^n = \left\{1+\frac{a\Delta}{a-1}\right\}^{-1}0^n.$

In the case of $a=-1$ an expression for A_n in terms of Bernoulli's numbers can be obtained.

For $\Sigma(-1)^x\phi(x) = (-1)^x(-e^D-1)^{-1}\phi(x)$, putting $a=-1$ in (23),

$$=(-1)^{x-1}(e^D+1)^{-1}\phi(x).$$

* In reality we may keep all terms up to $-\frac{B_{29}}{5^{31}}$, a quantity whose first significant figure is in the fourteenth decimal place.

Now $$\frac{1}{e^D+1}=\frac{1}{e^D-1}-\frac{2}{e^{2D}-1}$$

$$=\frac{1}{D}-\frac{1}{2}+\frac{B_1}{\lfloor 2}D-\ldots$$

$$-2\left\{\frac{1}{2D}-\frac{1}{2}+\frac{B_1}{\lfloor 2}2D-\ldots\right\}$$

$$=\frac{1}{2}-\frac{B_1}{\lfloor 2}(2^2-1)D+\frac{B_3}{\lfloor 4}(2^4-1)D^3-\ldots \qquad (25),$$

which determines the coefficients*.

10. *Expansion in inverse factorials.* The most general method of obtaining such expansions is by expressing the given function $\phi(x)$ in the form $\int_0^\infty e^{-xt}f(t)\,dt$. If we then write

$$e^{-t}=1-z, \text{ we get } \phi(x)=\int_0^1(1-z)^{x-1}f\left\{\log\left(\frac{1}{1-z}\right)\right\}dz.$$

$f\left\{\log\left(\frac{1}{1-z}\right)\right\}$ must now be expanded in some way in powers of z, and each term must be integrated separately by means of the formula

$$\int_0^1(1-z)^{x-1}z^m dz=\frac{\lfloor m}{x(x+1)\ldots(x+m)}.$$

By performing Σ on this we can expand in a similar way the more complicated form $\int_0^\infty \frac{e^{-t}-e^{-xt}}{e^{-t}-1}f(t)\,dt$. The most interesting cases are those in which $\phi(x)\equiv\log x$ or $\equiv\frac{1}{x^n}$ (see page 115).

The method is obviously very limited in its application. A paper on it by Schlömilch will be found in *Zeitschrift für*

* Compare (7), page 108. Ex. 12, page 85, is closely connected with the problem of this article.

Math. und Physik, IV. 390, and a review of this in Tortolini (*Annali*, 1859, 367) has sufficiently copious references to enable any one who desires it to follow out the subject. Stirling's formula—the earliest of the kind—is given in Ex. 11, page 30.

The very close connection that Factorials in general have with the Finite Calculus renders it worth while to give special attention to them, and to investigate in detail the laws of their transformations. For this purpose the student may consult a paper by Weierstrass (*Crelle*, LI. 1). Oettinger has also written on the subject (*Crelle*, XXXIII. and XXXVIII.), and Schläfli (*Crelle*, XLIII. and LXVII.). Ohm has an investigation into the connection between them and the Gamma-function (*Crelle*, XXXVI.), with a continuation on Factorials in general (*Crelle*, XXXIX.).

The papers on the subject of the Euler-Maclaurin Sum-formula are very numerous. Characteristic examples have been selected from them where it was possible, and placed, with references, in the accompanying Exercises.

By far the most important application of the principle of approximation is to the evaluation of Γx, or rather of $\log \Gamma x$ and its differential coefficients when x is very large. Raabe has two papers on this (*Crelle*, XXV. 146 and XXVIII. 10). See also Bauer (*Crelle*, LVII. 256) and Guderman (*Crelle*, XXIX. 209). Reference will be made to these papers when we consider Exact Theorems. See also a paper by Jeffery (*Quarterly Journal*, VI. 82) on the Derivatives of the Gamma-function. The constant C of Ex. 3 is of great importance in this theory. For its value, which has been calculated to a great number of decimal places, see *Crelle*, LX. 375.

Closely connected with the subject of differential coefficients of $\log \Gamma x$ is that of the summation of harmonic series $\left(\Sigma \frac{1}{\{a+(n-1)d\}^r}\right)$. On this see papers by Knar (Grunert, XLI. and XLIII.).

EXERCISES.

1. Find an expression for

$$\frac{1}{1^2}+\frac{1}{5^2}+\frac{1}{9^2}+\ldots, \text{ to } n \text{ terms},$$

and obtain an approximate value for the sum *ad infinitum*.

2. Find an approximate expression for $\Sigma \frac{1}{x^5}$ and also the value of

$$1+\frac{1}{2^5}+\frac{1}{3^5}+\ldots, \textit{ ad inf.},$$

to 10 places of decimals.

3. Find an approximate value of

$$\frac{3 \,.\, 5 \,.\, \dots\, (2x+1)}{2 \,.\, 4 \,.\, \dots\, 2x},$$

supposing x large but not infinite.

4. Find approximately $\Sigma \dfrac{1}{x^2 - a^2}$ and obtain an exact formula when a is an integral multiple of $\dfrac{1}{2}$.

5. Transform the series

$$\frac{1}{x^2+1} - \frac{1}{x^2+4} + \frac{1}{x^2+9} - \dots,$$

$$\frac{1}{x(x+1)(x+2)} - \frac{1}{(x+1)(x+2)(x+3)} + \dots,$$

into series of a more convergent character, and find an approximate value of the sum of each when $x = 5$, that is, correct to 6 places of decimals.

6. If $u_0 + u_1 x + u_2 x^2 + \dots = f(x)$, shew that

$$u_0 v_x + u_1 v_{x+1} + \dots = f(1) + f'(1)\, \Delta v_x + \frac{f''(1)}{1 \,.\, 2} \Delta^2 v_x + \dots$$

and apply this theorem to transform the series

$$x^{(-m)} + 2\frac{(x+1)^{(-m)}}{2} + 3\frac{(x+2)^{(-m)}}{2^2} + \dots$$

to one proceeding by factorials only.

7. Shew that

$$\frac{1}{z^2} + \frac{1}{(z+1)^2} + \frac{1}{(z+2)^2} + \dots$$

$$= \frac{1}{z} + \frac{1}{2 \,.\, z(z+1)} + \frac{1 \,.\, 2}{3 \,.\, z(z+1)(z+2)} + \dots.$$

8. Find the sum to n terms of the series

$$1+\frac{x^{(1)}}{z^{(1)}}+\frac{x^{(2)}}{z^{(2)}}+\dots,$$

and shew that its sum *ad inf.* is $\dfrac{z+1}{z-x+1}$.

9. Shew by the method given in the note to page 72, that

$$1^n+2^n+\dots+x^n=\frac{x^{n+1}}{n+1}+\frac{1}{2}x^n+\frac{\lfloor n}{\lfloor 2\,\lfloor n-2}B_1\frac{x^{n-1}}{n-1}+\dots,$$

where $B_{r-1}\equiv\dfrac{d^r}{dx^r}\left\{\dfrac{x}{1-e^{-x}}\right\}_{x=0}$ numerically.

[Schlömilch, *Grunert* x. 342.]

10. Shew that the sum of all the negative powers of all whole numbers (unity being in both cases excluded) is unity; if odd powers are excluded it is $\dfrac{3}{4}$.

11. Expand $\Sigma\dfrac{1}{(ax+b)^n}$ in terms of successive differences of $\log(ax+b)$ and deduce

$$\Sigma\cot x=C+\left\{\log\sin x-\frac{\Delta}{2}\log\sin x+\frac{\Delta^3}{3}\log\sin x-\dots\right\}.$$

[*Tortolini*, v. 281.]

12*. If $S_n=u_0+u_n+u_{2n}+\&c.$, *ad inf.*, shew that

$$S_n=\frac{1}{n}\left\{\sum_{r=0}^{r=\infty}u_r+\frac{n-1}{2}u_0-\frac{n^2-1}{12n}\Delta u_0+\frac{n^2-1}{24n}\Delta^2u_0-\dots\right\}.$$

13. Find $\Sigma\dfrac{1}{x^2}$ in factorials, and determine to 3 places of decimals the value of the constant when the first term is $\dfrac{1}{(3\frac{1}{2})^2}$.

If the Maclaurin Sum-formula had been used, to what degree of accuracy could we have obtained C?

* De Morgan (*Diff. Cal.* 554). Compare (27), page 54.

14. Shew that

$$\frac{1}{e^x-1}+\frac{1}{e^{2x}-1}+\dots,\ ad\ inf.,$$

$$=\frac{C-\log x}{x}+\frac{1}{4}-\frac{(B_1)^2}{2\,\underline{|2}}x-\frac{(B_3)^2}{4\,\underline{|4}}x^3-\dots$$

and apply this to the summation of Lambert's* series, viz.

$$\frac{x}{1-x}+\frac{x^2}{1-x^2}+\dots,\ \text{when } x \text{ is } \frac{1}{e} \text{ nearly.}$$

[*Zeitschrift*, VI. 407.]

15. Shew that

$$f(0)+f(1)+\dots\quad ad\ inf.$$

$$=\frac{1}{2}f(0)+\int_0^\infty \frac{e^{\pi t}+e^{-\pi t}}{e^{\pi t}-e^{-\pi t}}\frac{f(-\kappa t)-f(\kappa t)}{2\kappa}dt,$$

where $\kappa\equiv\sqrt{-1}$,

and deduce similar formulæ for the sums of the series

$$f(0)-f(1)+f(2)-\dots,$$

$$f(1)+f(3)+f(5)+\dots.$$

Find an analogous expression for the sum of the last mentioned to n terms.

16. Shew that

$$\frac{\sin x}{a+1}+\frac{\sin 2x}{a+2}+\frac{\sin 3x}{a+3}+\dots,\ ad\ inf.,$$

$$=\int_0^\infty \frac{e^{(\pi-x)t}-e^{-(\pi-x)t}}{e^{\pi t}-e^{-\pi t}}\frac{atdt}{a^2+t^2},$$

if x lie between π and $-\pi$.

[Schlömilch, *Crelle* XLII. 130.]

* On the application of the Maclaurin Sum-formula to this important series see also Curtze (*Annali Math.* I. 285).

CHAPTER VI.

BERNOULLI'S NUMBERS, AND FACTORIAL COEFFICIENTS.

1. THE celebrated series of numbers which we are about to notice were first discovered by James Bernoulli. They first presented themselves as connected with the coefficients of powers of x in the expression for the sum of the nth powers of the natural numbers, which we know is

$$1^n + 2^n \ldots + x^n = x^n + \Sigma x^n$$

$$= C + \frac{x^{n+1}}{n+1} + \frac{x^n}{2} + \frac{B_1}{\underline{|2}} n x^{n-1} - \frac{B_3}{\underline{|4}} n(n-1)(n-2)\, x^{n-3}$$

$$- \ldots \qquad (1),$$

or rather as the coefficient of x in the successive expressions when n was an even integer, and De Moivre pointed out that by taking this between limits 1 and 0 we obtain the formula

$$1^n = \frac{1}{n+1} + \frac{1}{2} + \frac{n}{\underline{|2}} B_1 - \frac{n(n-1)(n-2)}{\underline{|4}} B_3 + \ldots \qquad (2),$$

from which the numbers can be easily calculated in succession by taking $n = 2, 4, \ldots\ldots\ldots$

After the discovery of the Euler-Maclaurin formula [(6), page 90] the coefficients were shewn to be those of $\frac{1}{e^h - 1}$ from the application of it to Σe^{hx}, which gives

$$\frac{e^{hx}}{e^h - 1} = \Sigma e^{hx} = \int e^{hx} dx - \frac{1}{2} e^{hx} + \frac{B_1}{\underline{|2}} h e^{hx} - \ldots \qquad (3),$$

which gives

$$\frac{1}{e^h-1}=\frac{1}{h}-\frac{1}{2}+\frac{B_1}{\lfloor 2}h-\frac{B_3}{\lfloor 4}h^3+\dots \qquad (4).$$

2. Many other important expansions can be obtained by consideration of this identity.

Thus, for h write $2\theta\sqrt{-1}$; then, since

$$\frac{1}{e^{2\theta\sqrt{-1}}-1}=\frac{1}{2}\left\{\frac{e^{2\theta\sqrt{-1}}+1}{e^{2\theta\sqrt{-1}}-1}-1\right\}=\frac{1}{2\sqrt{-1}}\cot\theta-\frac{1}{2},$$

we at once obtain

$$\cot\theta=\frac{1}{\theta}-\frac{B_1}{\lfloor 2}2^2\theta-\frac{B_3}{\lfloor 4}2^4\theta^3-\dots \qquad (5).$$

Again $\operatorname{cosec}\theta=\cot\frac{\theta}{2}-\cot\theta$,

$$\therefore \operatorname{cosec}\theta=\frac{1}{\theta}+2(2-1)\frac{B_1}{\lfloor 2}\theta+2(2^3-1)\frac{B_3}{\lfloor 4}\theta^3+\cdots \qquad (6).$$

Similarly from $\cot\theta-2\cot 2\theta=\tan\theta$ we obtain

$$\tan\theta=\frac{2^2(2^2-1)}{\lfloor 2}B_1\theta+\frac{2^4(2^4-1)}{\lfloor 4}B_3\theta^3+\dots \qquad (7).$$

3. An expression for the values of the numbers of Bernoulli can be obtained from (5). For $\cot\theta=\frac{d}{d\theta}(\log\sin\theta)$ and

$$\log\sin\theta=\log\left\{\theta\left(1-\frac{\theta^2}{\pi^2}\right)\left(1-\frac{\theta^2}{2^2\pi^2}\right)\dots\right\}$$

$$=\log\theta+\log\left(1-\frac{\theta^2}{\pi^2}\right)+\cdots;$$

$$\therefore \cot\theta=\frac{1}{\theta}-\frac{2\theta}{\pi^2}\left(1-\frac{\theta^2}{\pi^2}\right)^{-1}-\frac{2\theta}{2^2\pi^2}\left(1-\frac{\theta^2}{2^2\pi^2}\right)^{-1}-\dots$$

$$=\frac{1}{\theta}-\frac{2\theta}{\pi^2}\left\{1+\frac{1}{2^2}+\frac{1}{3^2}+\dots\right\}$$

$$-\frac{2\theta^3}{\pi^4}\left\{1+\frac{1}{2^4}+\frac{1}{3^4}+\ldots\right\}$$

$$-\ \ldots \qquad (8).$$

Equating the coefficients of the same powers of θ in (5) and (8), we obtain

$$-\frac{2}{\pi^{2n}}\left(1+\frac{1}{2^{2n}}+\frac{1}{3^{2n}}+\ldots\right)=-\frac{B_{2n-1}}{\underline{|2n}}\,2^{2n},$$

$$\therefore B_{2n-1}=\frac{2\,\underline{|2n}}{(2\pi)^{2n}}\left\{1+\frac{1}{2^{2n}}+\frac{1}{3^{2n}}+\ldots\right\} \qquad (9)^*.$$

From this we see that the values of B_{2n-1} increase with very great rapidity, but those of $\frac{B_{2n-1}}{\underline{|2n}}$ ultimately approach to equality with those of a geometrical series whose common ratio is $\frac{1}{4\pi^2}$.

* A variation of (9), due I believe to Raabe (*Diff. und Int. Rechnung*, I. 412), depends on the following ingenious transformation :

$$S\equiv 1+\frac{1}{2^{2n}}+\frac{1}{3^{2n}}+\frac{1}{4^{2n}}+\ldots;$$

$$\therefore \frac{1}{2^{2n}}\cdot S=\frac{1}{2^{2n}}+\frac{1}{4^{2n}}+\ldots;$$

$$\therefore \left(1-\frac{1}{2^{2n}}\right)S=1+\frac{1}{3^{2n}}+\frac{1}{5^{2n}}+\ldots;$$

and all the terms of the form $\frac{1}{(2p)^{2n}}$ are removed. Proceeding as before

$$\left(1-\frac{1}{2^{2n}}\right)\left(1-\frac{1}{3^{2n}}\right)S=1+\frac{1}{5^{2n}}+\frac{1}{7^{2n}}+\ldots.$$

Thus we ultimately get

$$S=\frac{1}{\left(1-\frac{1}{2^{2n}}\right)\left(1-\frac{1}{3^{2n}}\right)\left(1-\frac{1}{5^{2n}}\right)\ldots}$$

where 2, 3, 5,... are the prime numbers taken in order. This formula would be of great use if we wished to obtain approximate values of B_n corresponding to large values of n, as it is well adapted for logarithmic computation.

4. If m be a positive integer and p be positive

$$\int_0^\infty e^{-px}x^m dx = \frac{m}{p}\int_0^\infty e^{-px}x^{m-1}dx = \dots = \frac{\underline{|m}}{p^{m+1}}.$$

Hence we can write (9) thus

$$B_{2n-1} = 4n\int_0^\infty x^{2n-1}\{e^{-2\pi x} + e^{-4\pi x} + \dots\}\,dx$$

$$= 4n\int_0^\infty \frac{x^{2n-1}}{e^{2\pi x}-1}\,dx \qquad (10)^*.$$

5. Euler was the first to call attention to a set of numbers closely analogous to those of Bernoulli. They appear in the coefficients of the powers of x when sec x is expanded. Thus

$$\sec x = 1 + \frac{E_2}{\underline{|2}}x^2 + \frac{E_4}{\underline{|4}}x^4 + \dots \qquad (11).$$

The identity $\sec x = \frac{d}{dx}\log\tan\left(\frac{\pi}{4} - \frac{x}{2}\right)$ will give, when treated as before,

$$E_{2n} = 2\frac{\underline{|2n}}{\left(\frac{\pi}{2}\right)^{2n+1}}\left\{1 - \frac{1}{3^{2n+1}} + \frac{1}{5^{2n+1}} - \dots\right\} \qquad (12),$$

while a consideration of the identity

$$\int_0^\infty \frac{\cos x\theta\,.\,d\theta}{e^{\frac{\pi}{2}\theta} + e^{-\frac{\pi}{2}\theta}} = \frac{1}{e^x + e^{-x}} \qquad (13)\dagger$$

will give

$$E_{2n} = 2\int_0^\infty \frac{\theta^{2n}\,.\,d\theta}{e^{\frac{\pi}{2}\theta} + e^{-\frac{\pi}{2}\theta}} \qquad (14),$$

formulæ analogous to (9) and (10), from which (12) may be deduced.

* Due to Plana (*Mém. de l'Acad. de Turin*, 1820).
† Schlömilch (*Grunert*, I. 361).

6. Owing to the importance of Bernoulli's and Euler's numbers a great many different formulæ have been investigated to facilitate their calculation. Most of these require them to be calculated successively from B_1 and E_2 onwards, and of these the most common for Bernoulli's numbers is (2). Others of a like kind may easily be obtained from the various expansions which involve them. Thus from (5), multiplying both sides by $\sin\theta$,

$$\cos\theta = \left(\theta - \frac{\theta^3}{|\underline{3}} + \dots\right)\left(\frac{1}{\theta} - \frac{B_1}{|\underline{2}}\,2^2\theta - \frac{B_3}{|\underline{4}}\,2^4\theta^3 - \dots\right),$$

and equating coefficients of θ^{2n} we obtain

$$\frac{(-1)^n}{|\underline{2n}} = -\left\{\frac{2^{2n}}{|\underline{2n}}\,B_{2n-1} - \frac{2^{2n-2}}{|\underline{3}\,|\underline{2n-2}}\,B_{2n-3} + \dots - \frac{(-1)^n}{|\underline{2n+1}}\right\} \qquad (15).$$

The simplest formulæ of this nature both for Bernoulli's and Euler's numbers are obtained at once from the original assumptions

$$\frac{t}{e^t-1} = 1 - \frac{t}{2} - \Sigma\,(-1)^n\frac{B_{n2-1}}{|\underline{2n}}\,t^{2n} \text{ and } \frac{1}{\cos t} = 1 + \Sigma\,\frac{E_{2n}}{|\underline{2n}}\,t^{2n}$$

by this method.

7. But direct expressions for the values of the numbers may be found. Thus

$$\frac{t}{e^t-1} = \frac{\log e^t}{e^t-1} = \frac{\log E}{E-1}\,e^{0\cdot t} \quad \text{(by Herschel's theorem)}$$

$$= \frac{\log(1+\Delta)}{\Delta}\,e^{0\cdot t}.$$

Hence, equating coefficients, we find

$$(-1)^{n+1}\frac{B_{2n-1}}{|\underline{2n}} = \frac{\log(1+\Delta)}{\Delta}\cdot\frac{0^{2n}}{|\underline{2n}};$$

$$\therefore\ B_{2n-1} = (-1)^{n+1}\left\{1 - \frac{\Delta}{2} + \frac{\Delta^2}{3} - \dots + \frac{\Delta^{2n}}{2n+1}\right\}0^{2n} \qquad (16),$$

and in like manner we obtain

$$0 = \left\{1 - \frac{\Delta}{2} + \frac{\Delta^2}{3} - \dots\right\} 0^{2n+1} \; (n > 0) \qquad (17).$$

8. These formulæ are capable of almost endless transformation. Thus, since $\Delta^{n-1} 0^{x-1} = \frac{\Delta^n 0^x}{n} - \Delta^n 0^{x-1}$ (Ex. 8, page 28), we can write (16) thus

$$B_{2n-1} = (-1)^{n+1} \left\{\left(\Delta - \frac{\Delta^2}{2^2} + \frac{\Delta^3}{3^2} - \dots\right) 0^{2n+1} - \left(\Delta - \frac{\Delta^2}{2} + \frac{\Delta^3}{3} - \dots\right) 0^{2n}\right\}$$

$$= (-1)^{n+1} \left(\Delta - \frac{\Delta^2}{2^2} + \frac{\Delta^3}{3^2} - \dots\right) 0^{2n+1} \qquad (18),$$

since the other term is

$$\log(1 + \Delta)\, 0^{2n} = D0^{2n} = 0.$$

9. A more general transformation by aid of the formula

$$f(\Delta)\, 0^n = Ef'(\Delta)\, 0^{n-1}$$

is as follows:

$$\{\log(1 + x\Delta)\}\, 0^n = \frac{xE}{1 + x\Delta} 0^{n-1} = \frac{x-1}{1 + x\Delta} 0^{n-1} \qquad (19).$$

Also

$$\{\log(1 + yE)\}\, 0 f(0) = yf(1) - \frac{y^2}{2} . 2f(2) + \dots$$

$$= yf(1) - y^2 f(2) + \dots$$

$$= \frac{yE}{1 + yE} f(0) = \left\{1 - \frac{1}{1 + yE}\right\} f(0)$$

$$= \frac{-1}{1 + yE} f(0) \qquad (20),$$

if $f(0) = 0$.

In (19) write E' for x and operate with each side on $f(0)$.

Then

$$\{\log(1+\Delta E')\}\,0^n f(0') = \frac{\Delta'}{1+\Delta E'}\,0^{n-1} f(0')$$

$$= -\{\log(1+\Delta E')\}\,0^{n-1}\,0'\Delta' f(0')$$

by (20), since $0^{n-1}\Delta' f(0') = 0$

$$= -\{\log(1+\Delta E')\}\,0^{n-1} f'(0'),$$

where $$f'(0') \equiv 0'\Delta' f(0').$$

Repeating this $n-1$ times we get

$$\{\log(1+\Delta E')\}\,0^n f(0') = (-1)^{n-1}\{\log(1+\Delta E')\}\,0 f^{n-1}(0')$$

$$= E' f^{n-1}(0') = [(x+1)\,\Delta\,(x+1)\,\Delta \ldots f(x+1)]_{x=0}.$$

This transformation has been given because it leads to a remarkable expression due to Bauer (*Crelle*, LVIII. 292) for Bernoulli's numbers.

Denote by Δ' the operating factor $(x+1)\,\Delta$, and write $\frac{1}{x}$ for $f(x)$ and $2n+1$ for n, and we obtain from (18)

$$B_{2n-1} = (-1)^{n-1}\{\log(1+\Delta E')\}\,\frac{0^{2n+1}}{0'} = \left[\Delta'^{2n}\left(\frac{1}{x+1}\right)\right]_{x=0} \qquad (21.)$$

Factorial Coefficients.

10. A series of numbers of great importance are those which form the coefficients of the powers of x when $x^{(n)}$ is expanded in powers of x. These usually go by the name of *factorial coefficients.*

It is evident by Maclaurin's Theorem that the coefficient of x^κ in the expansion of $x^{(n)}$ is $\dfrac{D^\kappa 0^{(n)}}{\lfloor\kappa}$ *. Although it is not

* Comparing (22) page 25, and (25) page 26, we see that $\dfrac{D^\kappa 0^{(n)}}{\lfloor n}$ is the coefficient of Δ^n in the expansion of $\{\log(1+\Delta)\}^\kappa$. That this is the case is

easy to obtain an expanded expression for this, it is very easy to calculate its successive values in a manner analogous to that used in Ch. II. Art. 13.

Let $C_\kappa^n \equiv$ numerical value of the coefficient of x^κ in the expansion of $x^{(n)}$. Then since $x^{(n+1)} = (x-n)\, x^{(n)}$, we obtain

$$C_\kappa^{n+1} = C_{\kappa-1}^n + n\, C_\kappa^n \qquad (22),$$

and we can thus calculate the values of C^{n+1} from those of C^n; and we know that the values of C^1 are 1, 0, 0, ...

11. Let us denote by C_κ^{-n} the numerical value of the coefficient of $\frac{1}{x^\kappa}$ in the expansion of $x^{(-n)}$ in negative powers of x, so that

$$x^{(-n)} = \frac{1}{x^n} - \frac{C_{n+1}^{-n}}{x^{n+1}} + \frac{C_{n+2}^{-n}}{x^{n+2}} - \ldots.$$

Then $x^{(-n)} = \frac{(-1)^{n-1}}{\lfloor n-1} \Delta^{n-1} \frac{1}{x} = \frac{(-1)^{n-1}}{\lfloor n-1} \left[\Delta^{n-1} \frac{1}{x+p}\right]_{p=0}$

(where Δ now refers to p alone)

$$= \frac{(-1)^{n-1}}{\lfloor n-1} \left[\Delta^{n-1} \left\{\frac{1}{x} - \frac{p}{x^2} + \frac{p^2}{x^3} - \ldots\right\}\right]_{p=0}$$

$$= \frac{(-1)^{n-1}}{\lfloor n-1} \left\{-\frac{\Delta^{n-1}0}{x^2} + \frac{\Delta^{n-1}0^2}{x^3} - \ldots\right\};$$

$$\therefore C_\kappa^{-n} = (-1)^{n-\kappa} \frac{\Delta^{n-1}0^{\kappa-1}}{\lfloor n-1} \qquad (23).$$

also evident from the following consideration:

$$\frac{D^\kappa 0^{(n)}}{\lfloor n} = \frac{D^{(n)} 0^\kappa}{\lfloor n} = \frac{1}{\lfloor n} \left\{z^n \frac{d^n}{dz^n} \overline{\log z|}^\kappa\right\}_{z=1}, \text{ putting } x = \log z$$

$$= \frac{1}{\lfloor n} \left\{\frac{d^n}{dz^n} \overline{\log(1+z)|}^\kappa\right\}_{z=0} = \text{coefficient of } z^n \text{ in the}$$

expansion of $\log(1+z)^\kappa$ by Maclaurin's theorem. Thus this expansion may be written

$$\frac{1}{\lfloor \kappa} \{\log(1+z)\}^\kappa = \frac{z^\kappa}{\lfloor \kappa} - \overset{\kappa+1}{C_\kappa} \cdot \frac{z^{\kappa+1}}{\lfloor \kappa+1} + \overset{\kappa+2}{C_\kappa} \frac{z^{\kappa+2}}{\lfloor \kappa+2} - \ldots$$

A formula analogous to (22) can also be obtained by means of Art. 13, Ch. II. This gives for numerical values

$$C_\kappa^{-n} = \frac{\Delta^{n-1}0^{\kappa-1}}{\lfloor n-1} = \frac{\Delta^{n-2}+\Delta^{n-1}}{\lfloor n-2} 0^{\kappa-2} = C_{\kappa-1}^{-(n-1)} + (n-1)\, C_{\kappa-1}^{-n} \qquad (24),$$

and thus from the values of $C_{\kappa-1}$ those of C_κ can be obtained. The values of C_1 are of course 1, 0, 0,...

12. Analogous series are those of the coefficients when x^n and x^{-n} are expanded in factorials.

By (5) page 11, we have

$$x^n = 0^n + \Delta 0^n . x + \frac{\Delta^2 0^n}{1.2} x^{(2)} + \ldots$$

$$= (-1)^{n-1} \left\{ C_{n+1}^{-2} x - C_{n+1}^{-3} x^{(2)} + \ldots \right\} \qquad (25),$$

following the notation of Art. 11.

Again in (25), page 26, put $u_x \equiv \frac{1}{x}$ and $\phi(D) \equiv D^{n-1}$, and we get after division by $(-1)^{n-1} \lfloor n-1$,

$$\frac{1}{x^n} = \frac{D^{n-1}0^{(n-1)}}{\lfloor n-1} x^{(-n)} - \frac{D^{n-1}0^{(n)}}{\lfloor n-1} x^{(-n-1)} + \ldots$$

$$= C_{n-1}^{n-1} x^{(-n)} - C_{n-1}^{n} x^{(-n-1)} + \ldots \qquad (27),$$

in the notation of Art. 10*.

* It will be seen that, as in the analogous case we could expand $\{\log(1+z)\}^\kappa$ in terms of C_κ^n, we can expand $(\epsilon-1)^\kappa$ in terms of $C_n^{-(\kappa+1)}$. In fact

$$(\epsilon^z - 1)^\kappa = z^\kappa + C_{\kappa+2}^{-(\kappa+1)} \frac{z^{\kappa+1}}{n+1} + C_{\kappa+3}^{-(\kappa+1)} \frac{z^{\kappa+2}}{(n+1)(n+2)} + \ldots \qquad (26)$$

where we have given $C_n^{-(\kappa+1)}$ its numerical value, disregarding its sign.

13. There is another class of properties of Bernoulli's numbers that has received some attention; these relate to their connection with the Theory of Numbers. Staudt's theorem will serve to illustrate the nature of these properties. It is that

$$B_{2n-1} = \text{integer} + (-1)^n \left(\frac{1}{2} + \Sigma \frac{1}{2m+1}\right)$$

where m is a divisor of n such that $2m+1$ is a prime number. Thus, taking $n=8$, we have (since the divisors of 8 are 1, 2, 4, 8)

$$B_{15} = \text{integer} + \left(\frac{1}{2} + \frac{1}{3} + \frac{1}{5} + \frac{1}{17}\right) = \text{integer} + 1\tfrac{47}{510}.$$

It will be found on reference to page 91 to be $7\frac{47}{510}$. Staudt's paper will be found in *Crelle* (XXI. 374), but a simpler demonstration of the above property has been given by Schläfli (*Quarterly Journal*, VI. 75). On this subject see papers by Kummer (*Crelle*, XL. XLI. LVI.). Staudt's theorem has also been given by Clausen.

14. To Raabe is due the invention of what he names the *Bernoulli-Function*, i.e. a function $F(x)$ given by

$$F(x) = 1^n + 2^n + \ldots + (x-1)^n$$

when x is an integer, and which is given generally by $\Delta F(x) = x^n$. He has also given the name *Euler-Function* to the analogous one that gives the sum of

$$1^n - 2^n + 3^n - \&c. + (2x-1)^n$$

when x is integral. See Brioschi (*Tortolini*, Series II. I. 260), in which there is a review of Raabe's paper (*Crelle*, XLII. 348) with copious references, and Kinkelin (*Crelle*, LVII. 122). See also a note by Cayley (*Quarterly Journal*, II. 198).

15. The most important papers on the subject of this Chapter are a series by Blissard (*Quarterly Journal*, Vols. IV.—IX.) under various titles. The demonstrations shew very strikingly the great power obtainable by the use of symbolical methods, which are here developed and applied to a much greater extent than in other papers on the subject. They include a most complete investigation into all the classes of numbers of which we have spoken in this Chapter; the results are too copious for any attempt to give them here, but Ex. 15 and 16 have been borrowed from them. The notation in the original differs from that here adopted. B_{2n} there denotes what is usually denoted by B_{2n-1}. See also two papers on $\Delta^n 0^m$ and its congeners by Horner (*Quarterly Journal*, IV.).

16. Attempts have been made to connect more closely Bernoulli's and Euler's Numbers, which we know already to have markedly similar properties. Scherk (*Crelle*, IV. 299) points out that, since $\tan\left(\frac{\pi}{4} + \frac{x}{2}\right) = \sec x + \tan x$, the expansion of this function in powers of x will have its coefficients depending alternately on each set of numbers }see (7) and (11), of this Chapter{. This idea has been taken up by others. Schlömilch (*Crelle*, XXXII. 360) has written a paper upon it. It enables us to represent both series by one expression, but there is no great advantage in doing so, as the expression referred to is very complicated. Another method is by finding the coefficient of x^n in the ex-

pansion of $\frac{1}{ae^x - 1}$, from which both series of numbers can be deduced by taking $a = \pm 1$ (Genocchi, *Tortolini*, Series I. Vol. III. 395).

17. Schlömilch has connected Bernoulli's numbers and factorial coefficients with the coefficients in the expansions of such quantities as $D^n f(\log x)$, $D^n\left(\frac{x}{f(x)}\right)$, &c. (*Grunert*, VIII. IX. XVI. XVIII.). Most of his analysis could be rendered simpler by the use of symbolical methods. This is usually the case in papers on this part of the subject, and the plan mentioned in the last Chapter has therefore been adhered to, of giving characteristic examples out of the various papers with references, instead of referring to them in the text. We must mention, in conclusion, that the numbers of Bernoulli as far as B_{31} have been calculated by Rothe, and will be found in *Crelle* (XX. 11).

EXERCISES.

1. Prove that

$$B_{2n-1} = (-1)^{n+1}\left\{\frac{\Delta 0^{2n+1}}{1^2} - \frac{\Delta^2 0^{2n+1}}{2^2} + \dots + \frac{\Delta^{2n+1} 0^{2n+1}}{(2n+1)^2}\right\}.$$

2. Prove that if n be an odd integer

$$\frac{1}{2} = nB_1 - \frac{n(n-1)(n-2)}{\underline{|3}} B_3 + \frac{n(n-1)(n-2)(n-3)(n-4)}{\underline{|5}} B_5, - \dots, \text{ to } n-1 \text{ terms.}$$

3. Obtain the formula of page 107, for determining successively Bernoulli's numbers, by differentiating the identity

$$t = -u + ue^t \text{ where } u = \frac{t}{e^t - 1}.$$

4. Shew that

$$B_n = \left(\frac{1}{2} - \frac{\Delta}{3} + \frac{\Delta^2}{4} - \dots\right) 1^n.$$

[Catalan, *Tortolini* 1859, 239.]

5. Shew that

$$B_{2n-1} = \frac{2x}{2^{2x}-1} \cdot \frac{(-1)^x}{2+\Delta} 0^{2x-1}.$$

6. Apply Herschel's Theorem to find an expression for a Bernoulli's number.

7. Demonstrate the following relation between the even Bernoulli's numbers:

$$\frac{2^{2n-1} B_{4n-1}}{\underline{|4n} \, \underline{|2}} - \frac{2^{2n-3} B_{4n-5}}{\underline{|4n-4} \, \underline{|6}} + \ldots + \frac{(-1)^n n}{\underline{|4n+2}} = 0.$$

[Knar, *Grunert*, XXVII. 455.]

8. Assuming the truth of the formula

$$\frac{e^x+1}{e^x-1} - \frac{2}{x} = 4 \int_0^\infty \frac{\sin xt}{e^{2\pi t}-1} \, dt,$$

deduce a value of B_{2n-1}.

9. Prove that the coefficient of θ^{2n} in the expansion of

$$\left(\frac{\theta}{\sin \theta}\right)^2 \text{ is equal to } \frac{2^{2n}(2n-1)}{\underline{|2n}} B_{2n-1}.$$

10. Express log sin x and log tan x in a series proceeding by powers of x by means of Bernoulli's numbers.

[Catalan, *Comptes Rendus*, LIV.]

11. Shew that the coefficient of

$$\frac{z^n}{\underline{|n}} \text{ in } \int_0^z \log(1-e^{-t}) \, dt - z \log z \text{ is } \frac{B_{n-2}}{n-1} \text{ numerically.}$$

12. Shew by Bernoulli's numbers or otherwise that

$$\frac{1^2}{1^2+1} \cdot \frac{2^2}{2^2+1} \cdot \frac{3^2}{3^2+1} \cdot \ldots \quad \textit{ad inf.} = \frac{2\pi}{e^\pi - e^{-\pi}}.$$

13. Prove that

$$1-\frac{B_1}{1.2}\pi^2+\frac{B_3}{1.2.3.4}\pi^4-\ldots=0.$$

14. Express the sums of the powers of numbers less than n and prime to it in series involving Bernoulli's numbers.

[Thacker, *Nouvelles Annales*, x. 324.]

15. If $\frac{t}{\epsilon^t-1}=1+P_1t+P_2t^2+\ldots$, shew that

$$\{(1+E)^n-E^n\}P_0=0,$$

$$\left(E+\frac{1}{2}\right)^{2n+1}P_0=0,$$

$$(1+x)^EP_0=\frac{\log(1+x)}{x}.$$

16. Shew, in the notation of the last question, that

$$\frac{E^{n+1}(E-1)(E-2)}{2}P_0=(-1)^n\left(\frac{\Delta}{4}-\frac{\Delta^2}{5}+\ldots\right)0^{n+1}.$$

17. Shew that

$$\frac{\sin x}{1^{2r+1}}-\frac{\sin 2x}{2^{2r+1}}+\frac{\sin 3x}{3^{2r+1}}-\ldots$$

$$=(-1)^r\left\{\frac{1}{2}\frac{x^{2r+1}}{\underline{|2r+1}}-B_1\frac{x^{2r-1}}{\underline{|2r-1}}+B_3\frac{x^{2r-3}}{\underline{|2r-3}}-\ldots\right\},$$

where $$B_n\equiv 1-\frac{1}{2^{2r}}+\frac{1}{3^{2r}}-\ldots$$

and hence find the sum of such a series in terms of Bernoulli's numbers.

[Dienger, *Crelle*, XXXIV. 91.]

18. Shew that*

$$1-\frac{1}{3^3}+\frac{1}{5^3}-\ldots=\frac{\pi^3}{32},$$

$$1-\frac{1}{3^5}+\frac{1}{5^5}-\ldots=\frac{5\pi^5}{1536}.$$

* Many similar summations will be found in a paper by Tchebechef [*Liouville*, XVI. 337].

19. If $S_x^n \equiv 1^n + 2^n + \ldots + x^n$, shew that

$$S_x^{\,n} = xS_{x-1}^n - \sum_{r=1}^{r=x-1} S_r^{\,n-1}.$$

[Eisenlohe, *Crelle*, XXVIII. 193.]

20. In the notation of Art. 14, page 116, shew that $F(x)$ is a rational and integral function of $x - \frac{1}{2}$, and cannot contain both odd and even powers of the same.

[Bertrand, *Diff. Cal.* 350.]

21. Shew how the method contained in the note on page 109 could be made to give us the actual values of the numbers of Bernoulli by application of Staudt's Theorem.

22. Apply the formula,

$$\cos\frac{x}{4} + \sin\frac{x}{4} = (1+x)\left(1 - \frac{1}{3}x\right)\left(1 + \frac{1}{5}x\right). \ \ldots$$

to demonstrate (12), page 110.

[Stern, *Crelle*, XXVI. 88.]

23. If $F(n) \equiv \dfrac{2\kappa}{1-\kappa}\left[1 - \left\{1 + \dfrac{1-\kappa}{2}\Delta\right\}^{-1}\right] 0^n$, where

$$\kappa \equiv \sqrt{-1},$$

shew that

$$E_{2n} = F(2n), \text{ and } \frac{2^{2n}(2^{2n}-1)}{2n} B_{2n-1} = \kappa F(2n-1) \text{ numerically.}$$

[Schlömilch, *Crelle*, XXXII. 360.]

24. Shew that

$$\frac{\Delta^m 0^{m+p}}{\underline{|m}} = \Sigma\,[(m+1)\,\Sigma\,\{(m+2) \ldots\ldots \Sigma\,(m+p)\}].$$

[Hargreave, *Quarterly Journal*, VIII. 26.]

25. *Shew that $f(D)\,0^{(n)} = \underline{|n}\,\dfrac{f(D)}{\Delta^{n-1}}\,0 = \dfrac{d^{n-1}f(D)}{d\Delta^{n-1}}\,0.$

*Prove that $\dfrac{\Delta^r}{\underline{|r}}\{0+(n-r)\}^{s+r}$ expresses the sum of all the homogeneous products of s dimensions which can be formed of the $r+1$ consecutive numbers $n, n-1, \ldots n-r$.

26. Express $x^{(m)} \times x^{(n)}$ in factorials.

[Elphinstone, *Quarterly Journal*, II. 254.]

27. If $\log(1+x) = A_1 x + A_2 x^{(2)} + A_3 x^{(3)} + \ldots,$

shew that $$A_4 = \frac{1}{4}\left\{\log\frac{3}{4} + \frac{1}{6}\log 5\right\}.$$

28. If $K_r^m \equiv$ number of combinations of m things r together with repetitions,

$C_r^m \equiv$ number of combinations of m things r together without repetitions,

then $K_r^m = \dfrac{\Delta^m 0^{m+r}}{\underline{|m}}$, and C_r^m is obtained by writing $-(m+1)$ for m in the expanded expression for K_r^m.

[Wasmund, *Grunert*, XXXIV. 440.]

29. Shew that in the notation of Art. 10

$$C_1^n - C_2^n + C_3^n - \ldots = 0,$$

and $B_1 C_2^n - B_3 C_4^n + \ldots = \dfrac{\underline{|n-1}}{2}\cdot\dfrac{n-1}{n+1}.$

[*Grunert*, IX. 333.]

30. Shew that

$$x(x-1).\ \ldots\ (x-m+1) = \frac{2\,\underline{|m}}{\pi}\int_0^\pi 2\left(\cos\frac{z}{2}\right)^x \cos\frac{xz}{2}\cos mz\,dz;$$

and find from this an expression for the coefficients of the powers of x in the expanded factorial $x^{(m)}$ in the form of a definite integral.

[*Grunert*, XI. 447.]

* Jeffery (*Quarterly Journal*, IV. 364).

31. Deduce (26) page 115, from (21) page 24.

32. Shew that $C_4^{-3} = 3$, and $C_4^{6} = 85$.

33. Shew that if $\kappa < n$ the coefficient of

$$x^\kappa \text{ in } \left(\frac{x}{e^x - 1}\right)^n \text{ is } \frac{(-1)^\kappa C_{n-\kappa}^n}{(n-1)(n-2)\ldots(n-\kappa)},$$

in the notation of Art. 10.

[Schlömilch, *Grunert*, XVIII. 315.]

34. Shew that (with the notation of (21), page 113)

$$\left[\Delta'^n \left(\frac{1}{r}\right)\right]_{r=3} = \frac{1}{3}\left[\left\{\Delta'^n + \left(1 + \frac{1}{2}\right)\Delta'^{n+1} + \frac{1}{1 \,.\, 2}\Delta'^{n+2}\right\}\left(\frac{1}{r}\right)\right]_{r=1},$$

and find the general formula for $r = \kappa$.

Shew that

$$2\left[\Delta'^{2n}\left(\frac{1}{r}\right)\right]_{r=2} = \left(-1\right)^{n-1} B_{2n-1}.$$

35. If $x\dfrac{d}{dx} \equiv D_1$, shew that

$$D_1^{n+1} f(x) \equiv xf'(x) + \frac{\Delta 1^n}{1} x^2 f''(x) + \frac{\Delta^2 1^n}{1 \,.\, 2} x^3 f'''(x) + \ldots.$$

36. Find expressions for Bernoulli's numbers and Factorial-coefficients in the form of determinants.

[*Tortolini*, Series II. VII. 19.]

CHAPTER VII.

CONVERGENCY AND DIVERGENCY OF SERIES.

1. A SERIES is said to be convergent or divergent according as the sum of its first n terms approaches or does not approach to a finite limit when n is indefinitely increased.

This definition leads us to distinguish between the convergency of a series and the convergency of the *terms* of a series. The successive terms of the series

$$1 + \frac{1}{2} + \frac{1}{3} + \frac{1}{4} + \frac{1}{5} + \ldots$$

converge to the limit 0, but it will be shewn that the sum of n of those terms tends to become infinite with n.

On the other hand, the geometrical series

$$1 + \frac{1}{2} + \frac{1}{4} + \frac{1}{8} + \frac{1}{16} + \ldots$$

is convergent both as respects its terms and as respects the sum of its terms.

2. Three cases present themselves. 1st. That in which the terms of a series are all of the same or are ultimately all of the same sign. 2ndly. That in which they are, or ultimately become, alternately positive and negative. 3rdly. That in which they are of variable sign (though not alternately positive and negative) owing to the presence of a periodic quantity as a factor in the general term. The first case we propose, on account of the greater difficulty of its theory, to consider last.

3. PROP. 1. *A series whose terms diminish in absolute value, and are, or end with becoming, alternately positive and negative, is convergent.*

Let $u_1 - u_2 + u_3 - u_4 + \dots$ be the proposed series or its terminal portion, the part which it follows being in the latter case supposed finite. Then, writing it in the successive forms

$$u_1 - u_2 + (u_3 - u_4) + (u_5 - u_6) + \cdots \qquad (1),$$

$$u_1 - (u_2 - u_3) - (u_4 - u_5) - \dots \qquad (2),$$

and observing that $u_1 - u_2$, $u_2 - u_3$, ... are by hypothesis positive, we see that the sum of the series is greater than $u_1 - u_2$ and less than u_1. The series is therefore convergent.

Ex. Thus the series

$$1 - \frac{1}{2} + \frac{1}{3} - \frac{1}{4} + \frac{1}{5} - \dots \textit{ ad inf.}$$

tends to a limit which is less than 1 and greater than $\frac{1}{2}$*.

4. PROP. II. *A series whose n^{th} term is of the form $u_n \sin n\theta$ (where θ is not zero or an integral multiple of 2π) will converge if, for large values of n, u_n retains the same sign, continually diminishes as n increases, and ultimately vanishes.*

Suppose u_n to retain its sign and to diminish continually as n increases after the term u_a. Let

$$S = u_a \sin a\theta + u_{a+1} \sin (a+1)\theta + \dots \qquad (3);$$

* Although the above demonstration is quite rigorous, still such series present many analogies with divergent series and require careful treatment. For instance, in a convergent series where all the terms have the same sign, the order in which the terms are written does not affect the sum of the series. But in the given case, if we write the series thus,

$$\left(1 + \frac{1}{3}\right) - \frac{1}{2} + \left(\frac{1}{5} + \frac{1}{7}\right) - \frac{1}{4} + \dots$$

in which form it is equally convergent, we find that its value lies between $\frac{5}{6}$ and $\frac{4}{3}$ while that of the original series lies between $1 - \frac{1}{2}$ and $1 - \frac{1}{2} + \frac{1}{3}$, *i.e.* between $\frac{1}{2}$ and $\frac{5}{6}$.

$$\therefore\ 2\sin\frac{\theta}{2}S = u_a\left\{\cos\left(a-\frac{1}{2}\right)\theta - \cos\left(a+\frac{1}{2}\right)\theta\right\}$$

$$+ u_{a+1}\left\{\cos\left(a+\frac{1}{2}\right)\theta - \cos\left(a+\frac{3}{2}\right)\theta\right\} + \dots$$

$$= u_a\cos\left(a-\frac{1}{2}\right)\theta + (u_{a+1}-u_a)\cos\left(a+\frac{1}{2}\right)\theta$$

$$+ (u_{a+2}-u_{a+1})\cos\left(a+\frac{3}{2}\right)\theta + \dots.$$

Now $u_{a+1}-u_a$, $u_{a+2}-u_{a+1}$, ... are all negative, hence

$$2\sin\frac{\theta}{2}S - u_a\cos\left(a-\frac{1}{2}\right)\theta < (u_{a+1}-u_a) + (u_{a+2}-u_{a+1}) + \dots$$

numerically,

$$\text{or} < u_\infty - u_a;\ \therefore < -u_a,\ \text{since } u_\infty = 0.$$

Hence the series is convergent unless $\sin\frac{\theta}{2}$ be zero, i.e. unless θ be zero or an integral multiple of 2π*.

An exactly similar demonstration will prove the proposition for the case in which the nth term is $u_n \sin(n\theta - \beta)$.

Ex. The series

$$\sin\theta + \frac{\sin 2\theta}{2} + \frac{\sin 3\theta}{3} + \dots$$

is convergent unless θ be zero or a multiple of 2π. This is the case although, as we shall see, the series

$$1 + \frac{1}{2} + \frac{1}{3} + \dots \text{ is divergent.}$$

5. The theory of the convergency and divergency of series whose terms are ultimately of one sign and at the same time converge to the limit 0, will occupy the remainder of this chapter and will be developed in the following order. 1st. A

* Malmstén (*Grunert*, VI. 38). A more general proposition is given by Chartier (*Liouville*, XVIII. 21).

fundamental proposition, due to Cauchy, which makes the test of convergency to consist in a process of integration, will be established. 2ndly. Certain direct consequences of that proposition relating to particular classes of series, including the geometrical, will be deduced. 3rdly. Upon those consequences, and upon a certain extension of the algebraical theory of *degree* which has been developed in the writings of Professor De Morgan and of M. Bertrand, a system of criteria general in application will be founded. It may be added that the first and most important of the criteria in question, to which indeed the others are properly supplemental, being founded upon the known properties of geometrical series, might be proved without the aid of Cauchy's proposition; but for the sake of unity it has been thought proper to exhibit the different parts of the system in their natural relation.

Fundamental Proposition.

6. PROP. III. *If the function $\phi(x)$ be positive in sign but diminishing in value as x varies continuously from a to ∞, then the series*

$$\phi(a) + \phi(a+1) + \phi(a+2) + \ldots \textit{ ad inf.} \qquad (4)$$

will be convergent or divergent according as $\int_a^\infty \phi(x)\,dx$ is finite or infinite.

For, since $\phi(x)$ diminishes from $x = a$ to $x = a + 1$, and again from $x = a + 1$ to $x = a + 2$, ..., we have

$$\int_a^{a+1} \phi(x)\,dx < \phi(a),$$

$$\int_{a+1}^{a+2} \phi(x)\,dx < \phi(a+1),$$

and so on, *ad inf.* Adding these inequations together, we have

$$\int_a^\infty \phi(x)\,dx < \phi(a) + \phi(a+1) + \ldots \textit{ ad inf.} \qquad (5).$$

Again, by the same reasoning,

$$\int_a^{a+1} \phi(x)\,dx > \phi(a+1),$$

$$\int_{a+1}^{a+2} \phi(x)\,dx > \phi(a+2),$$

and so on. Again adding, we have

$$\int_a^{\infty} \phi(x)\,dx > \phi(a+1) + \phi(a+2) + \ \dots \qquad (6).$$

Thus the integral $\int_a^{\infty} \phi(x)\,dx$, being intermediate in value between the two series

$$\phi(a) + \phi(a+1) + \phi(a+2) + \dots$$
$$\phi(a+1) + \phi(a+2) + \dots$$

which differ by $\phi(a)$, will differ from the former series by a quantity less than $\phi(a)$, therefore by a finite quantity. Thus the series and the integral are finite or infinite together.

COR. *If in the inequation* (6) *we change* a *into* $a-1$, *and compare the result with* (5), *it will appear that the series*

$$\phi(a) + \phi(a+1) + \phi(a+2) + \ \dots \ ad\ inf.$$

has for its inferior and superior limits

$$\int_a^{\infty} \phi(x)\,dx, \text{ and} \int_{a-1}^{\infty} \phi(x)\,dx. \qquad (7).$$

7. The application of the above proposition will be sufficiently explained in the two following examples relating to geometrical series and to the other classes of series involved in the demonstration of the final system of criteria referred to in Art. 5.

Ex. 1. The geometrical series

$$1 + h + h^2 + h^3 + \ \dots \ ad\ inf.$$

is convergent if $h < 1$, divergent if $h \geqq 1$.

The general term is h^x, the value of x in the first term being 0, so that the test of convergency is simply whether $\int_0^\infty h^x dx$ is infinite or not. Now

$$\int_0^x h^x dx = \frac{h^x - 1}{\log h}.$$

If $h > 1$ this expression becomes infinite with x and the series is divergent. If $h < 1$ the expression assumes the finite value $\frac{-1}{\log h}$. The series is therefore convergent.

If $h = 1$ the expression becomes indeterminate, but, proceeding in the usual way, assumes the limiting form xh^x which becomes infinite with x. Here then the series is divergent.

Ex. 2. The successive series

$$\left.\begin{array}{c} \dfrac{1}{a^m} + \dfrac{1}{(a+1)^m} + \dfrac{1}{(a+2)^m} + \dots \\ \dfrac{1}{a(\log a)^m} + \dfrac{1}{(a+1)\{\log(a+1)\}^m} + \dots \\ \dfrac{1}{a\log a(\log\log a)^m} + \dfrac{1}{(a+1)\log(a+1)\{\log\log(a+1)\}^m} + \dots \\ \dots\dots\dots\dots\dots\dots\dots\dots\dots\dots \end{array}\right\} \quad (8)^*,$$

a being positive, are convergent if $m > 1$, and divergent if $m \leqq 1$.

The determining integrals are

$$\int_a^\infty \frac{dx}{x^m}, \quad \int_a^\infty \frac{dx}{x(\log x)^m}, \quad \int_a^\infty \frac{dx}{x\log x(\log\log x)^m}, \dots\dots$$

* The convergency of these series can be investigated without the use of the Integral Calculus. See Todhunter's *Algebra* (Miscellaneous Theorems), or Malmstén (*Grunert*, VIII. 419).

and their values, except when m is equal to 1, are

$$\frac{x^{1-m}-a^{1-m}}{1-m}, \quad \frac{(\log x)^{1-m}-(\log a)^{1-m}}{1-m}, \quad \frac{(\log\log x)^{1-m}-(\log\log a)^{1-m}}{1-m} \ldots$$

in which $x=\infty$. All these expressions are infinite if m be less than 1, and finite if m be greater than 1. If $m=1$ the integrals assume the forms

$$\log x - \log a, \ \log\log x - \log\log a, \ \log\log\log x - \log\log\log a \ \ldots$$

and still become infinite with x. Thus the series are convergent if $m > 1$ and divergent if $m \leqq 1$.

Perhaps there is no other mode so satisfactory for establishing the convergency or divergency of a series as the direct application of Cauchy's proposition, when the integration which it involves is possible. But, as this is not always the case, the construction of a system of derived rules not involving a process of integration becomes important. To this object we now proceed.

First derived Criterion.

8. PROP. IV. *The series $u_0 + u_1 + u_2 + \ldots$ ad inf., all whose terms are supposed positive, is convergent or divergent according as the ratio $\frac{u_{x+1}}{u_x}$ tends, when x is indefinitely increased, to a limiting value less or greater than unity.*

Let h be that limiting value; and first let h be less than 1, and let k be some positive quantity so small that $h+k$ shall also be less than 1. Then as $\frac{u_{x+1}}{u_x}$ tends to the limit h, it is possible to give to x some value n so large, yet finite, that for that value and for all superior values of x the ratio $\frac{u_{x+1}}{u_x}$ shall lie within the limits $h+k$ and $h-k$. Hence if, beginning with the particular value of x in question, we construct the three series

$$\left.\begin{array}{l} u_n + (h+k)\, u_n + (h+k)^2\, u_n + \dots \\ u_n + u_{n+1} \qquad + \qquad u_{n+2} \qquad + \dots \\ u_n + (h-k)\, u_n + (h-k)^2\, u_n + \dots \end{array}\right\} \quad (9),$$

each term after the first in the second series will be intermediate in value between the corresponding terms in the first and third series, and therefore the second series will be intermediate in value between

$$\frac{u_n}{1-(h+k)} \text{ and } \frac{u_n}{1-(h-k)},$$

which are the finite values of the first and third series. And therefore the given series is convergent.

On the other hand, if h be greater than unity, then, giving to k some small positive value such that $h-k$ shall also exceed unity, it will be possible to give to x some value n so large, yet finite, that for that and all superior values of x, $\frac{u_{x+1}}{u_x}$ shall lie between $h+k$ and $h-k$. Here then still each term after the first in the second series will be intermediate between the corresponding terms of the first and third series. But $h+k$ and $h-k$ being both greater than unity, both the latter series are divergent (Ex. 1). Hence the second or given series is divergent also.

Ex. 3. The series $1+t+\frac{t^2}{1.2}+\frac{t^3}{1.2.3}+\dots$, derived from the expansion of ϵ^t, is convergent for all values of t.

For if

$$u_x = \frac{t^x}{1.2\dots x}, \quad u_{x+1} = \frac{t^{x+1}}{1.2\dots(x+1)},$$

then

$$\frac{u_{x+1}}{u_x} = \frac{t}{x+1},$$

and this tends to 0 as x tends to infinity.

Ex. 4. The series

$$1+\frac{a}{b}t+\frac{a(a+1)}{b(b+1)}t^2+\frac{a(a+1)(a+2)}{b(b+1)(b+2)}t^3+\dots$$

is convergent or divergent according as t is less or greater than unity.

Here
$$u_x=\frac{a(a+1)(a+2)\dots(a+x-1)}{b(b+1)(b+2)\dots(b+x-1)}t^x.$$

Therefore
$$\frac{u_{x+1}}{u_x}=\frac{a+x}{b+x}t,$$

and this tends, x being indefinitely increased, to the limit t. Accordingly therefore as t is less or greater than unity, the series is convergent or divergent.

If $t=1$ the rule fails. Nor would it be easy to apply directly Cauchy's test to this case, because of the indefinite number of factors involved in the expression of the general term of the series. We proceed, therefore, to establish the supplemental criteria referred to in Art. 5.

Supplemental Criteria.

9. Let the series under consideration be

$$u_a+u_{a+1}+u_{a+2}+u_{a+3}+\dots \textit{ad inf.} \qquad (10),$$

the general term u_x being supposed positive and diminishing in value from $x=a$ to $x=$ infinity. The above form is adopted as before to represent the terminal, and by hypothesis positive, portion of series whose terms do not necessarily begin with being positive; since it is upon the character of the terminal portion that the convergency or divergency of the series depends.

It is evident that the series (10) will be convergent if its terms become ultimately less than the corresponding terms of a known convergent series, and that it will be divergent if its terms become ultimately greater than the corresponding terms of a known divergent series.

Compare then the above series whose general term is u_x with the first series in (8), Ex. 2, whose general term is $\frac{1}{x^m}$. Then a condition of convergency is

$$u_x < \frac{1}{x^m},$$

m being greater than unity, and x being indefinitely increased.

Hence we find

$$x^m < \frac{1}{u_x};$$

$$\therefore m \log x < \log \frac{1}{u_x};$$

$$m < \frac{\log \frac{1}{u_x}}{\log x},$$

and since m is greater than unity

$$\frac{\log \frac{1}{u_x}}{\log x} > 1.$$

On the other hand, there is divergency if

$$u_x > \frac{1}{x^m},$$

x being indefinitely increased, and m being equal to or less than 1. But this gives

$$m > \frac{\log \frac{1}{u_x}}{\log x},$$

and therefore

$$\frac{\log \frac{1}{u_x}}{\log x} < 1.$$

It appears therefore that the series *is convergent or divergent according as, x being indefinitely increased, the function* $\dfrac{\log \frac{1}{u_x}}{\log x}$ *approaches a limit greater or less than unity.*

But the limit being unity, and the above test failing, let the comparison be made with the second of the series in (8). For convergency, we then have as the limiting equation,

$$u_x < \frac{1}{x(\log x)^m},$$

m being greater than unity. Hence we find, by proceeding as before,

$$\frac{\log \frac{1}{xu_x}}{\log \log x} > 1.$$

And deducing in like manner the condition of divergency, we conclude that *the series is convergent or divergent according as, x being indefinitely increased, the function* $\dfrac{\log \frac{1}{xu_x}}{\log \log x}$ *tends to a limit greater or less than unity.*

Should the limit be unity, we must have recourse to the third series of (8), the resulting test being that *the proposed series is convergent or divergent according as, x being indefinitely increased, the function* $\dfrac{\log \frac{1}{x \log xu_x}}{\log \log \log x}$ *tends to a limit greater or less than unity.*

The forms of the functions involved in the succeeding tests, *ad inf.*, are now obvious. Practically, we are directed to construct the successive functions,

$$\frac{l\frac{1}{u_x}}{lx}, \quad \frac{l\frac{1}{xu_x}}{llx}, \quad \frac{l\frac{1}{xlxu_x}}{lllx}, \quad \frac{l\frac{1}{xlxllxu_x}}{llllx}, \ \ldots \qquad \text{(A)},$$

and the first of these which tends, as x is indefinitely increased to a limit greater or less than unity, determines the series to be convergent or divergent.

The criteria may be presented in another form. For representing $\frac{1}{u_x}$ by $\phi(x)$, and applying to each of the functions in (A), the rule for indeterminate functions of the form $\frac{\infty}{\infty}$, we have

$$\frac{l\phi(x)}{lx} = \frac{\phi'(x)}{\phi(x)} \div \frac{1}{x} = \frac{x\phi'(x)}{\phi(x)},$$

$$\frac{l\dfrac{\phi(x)}{x}}{llx} = \left\{\frac{\phi'(x)}{\phi(x)} - \frac{1}{x}\right\} \div \frac{1}{x\log x}$$

$$= \log x \left\{x\frac{\phi'(x)}{\phi(x)} - 1\right\},$$

and so on. Thus the system of functions (A) is replaced by the system

$$\frac{x\phi'(x)}{\phi(x)},\quad lx\left\{\frac{x\phi'(x)}{\phi(x)} - 1\right\},$$

$$llx\left[lx\left\{x\frac{\phi'(x)}{\phi(x)} - 1\right\} - 1\right],\ \ldots \qquad \text{(B)}.$$

It was virtually under this form that the system of functions was originally presented by Prof. De Morgan, (*Differential Calculus*, pp. 325—7). The law of formation is as follows. If P_n represent the nth function, then

$$P_{n+1} = l^n x\,(P_n - 1) \qquad (11).$$

10. There exists yet another and equivalent system of determining functions which in particular cases possesses great advantages over the two above noted. It is obtained by substituting in Prof. De Morgan's forms $\frac{u_x}{u_{x+1}} - 1$ for $\frac{\phi'(x)}{\phi(x)}$. The lawfulness of this substitution may be established as follows.

Since $u_x = \frac{1}{\phi(x)}$, we have

$$\frac{u_x}{u_{x+1}} - 1 = \frac{\phi(x+1)}{\phi(x)} - 1$$

$$= \frac{\phi(x+1) - \phi(x)}{\phi(x)}$$

$$= \frac{\phi'(x+\theta)}{\phi(x)}$$

(θ being some quantity between 0 and 1)

$$= \frac{\phi'(x)}{\phi(x)} \frac{\phi'(x+\theta)}{\phi'(x)} \qquad (12).$$

Now $\frac{\phi'(x+\theta)}{\phi'(x)}$ has unity for its limiting value; for, $\phi(x)$ tends to become infinite as x is indefinitely increased, and therefore $\frac{\phi(x+\theta)}{\phi(x)}$ assumes the form $\frac{\infty}{\infty}$; therefore

$$\frac{\phi(x+\theta)}{\phi(x)} = \frac{\phi'(x+\theta)}{\phi'(x)}.$$

And thus the second member has for its limits $\frac{\phi(x)}{\phi(x)}$ and $\frac{\phi(x+1)}{\phi(x)}$, i.e. 1 and $\frac{\phi(x+1)}{\phi(x)}$; or in other words tends to the limit 1. Thus (12) becomes

$$\frac{u_x}{u_{x+1}} - 1 = \frac{\phi'(x)}{\phi(x)}.$$

Substituting therefore in (B), we obtain the system of functions

$$x\left(\frac{u_x}{u_{x+1}} - 1\right), \quad lx\left\{x\left(\frac{u_x}{u_{x+1}} - 1\right) - 1\right\},$$

$$llx\left[lx\left\{x\left(\frac{u_x}{u_{x+1}} - 1\right) - 1\right\} - 1\right] \dots \qquad \text{(C)},$$

the law of formation being still $P_{n+1} = l^n x(P_n - 1)$.

11. The extension of the theory of *degree* referred to in Art. 5 is involved in the demonstration of the above criteria. When two functions of x are, in the ordinary sense of the term, of the same degree, i.e. when they respectively involve the same highest powers of x, they tend, x being indefinitely increased, to a ratio which is finite yet not equal to 0; viz. to the ratio of the respective coefficients of that highest power. Now let the converse of this proposition be assumed as the *definition* of equality of degree, i.e. let *any* two functions of x be said to be of the same degree when the ratio between them tends, x being indefinitely increased, to a finite limit which is not equal to 0. Then are the several functions

$$x(lx)^m, \quad xlx(llx)^m, \quad \ldots,$$

with which $\frac{1}{u_x}$ or $\phi(x)$ is successively compared in the demonstrations of the successive criteria, so many interpositions of degree between x and x^{1+a}, however small a may be. For x being indefinitely increased, we have

$$\lim \frac{x(lx)^m}{x} = \infty, \quad \lim \frac{x(lx)^m}{x^{1+a}} = 0,$$

$$\lim \frac{xlx(llx)^m}{xlx} = \infty, \quad \lim \frac{xlx(llx)^m}{x(lx)^{1+a}} = 0,$$

so that, according to the definition, $x(lx)^m$ is intermediate in degree between x and x^{1+a}, $xlx(llx)^m$ between xlx and $x(lx)^{1+a}, \ldots$. And thus each failing case, arising from the supposition of $m = 1$, is met by the introduction of a new function.

It may be noted in conclusion that the first criterion of the system (A) was originally demonstrated by Cauchy, and the first of the system (C) by Raabe (*Crelle*, Vol. IX.). Bertrand*, to whom the comparison of the three systems is due, has demonstrated that if one of the criteria should fail from the absence of a *definite* limit, the succeeding criteria will also fail in the same way. The possibility of their continued failure through the continued reproduction of the definite limit 1, is a question which has indeed been noticed but has scarcely been discussed.

* Liouville's *Journal*, Tom. VII. p. 35.

12. The results of the above inquiry may be collected into the following rule.

RULE. *Determine first the limiting value of the function* $\frac{u_{x+1}}{u_x}$. *According as this is less or greater than unity the series is convergent or divergent.*

But if that limiting value be unity, seek the limiting values of whichsoever is most convenient of the three systems of functions (A), (B), (C). *According as, in the system chosen, the first function whose limiting value is not unity, assumes a limiting value greater or less than unity, the series is convergent or divergent.*

Ex. 5. Let the given series be

$$1 + \frac{1}{2^{\frac{3}{2}}} + \frac{1}{3^{\frac{4}{3}}} + \frac{1}{4^{\frac{5}{4}}} + \dots \qquad (13).$$

Here
$$u_x = \frac{1}{x^{\frac{x+1}{x}}}, \text{ therefore,}$$

$$\frac{u_{x+1}}{u_x} = \frac{x^{\frac{x+1}{x}}}{(x+1)^{\frac{x+2}{x+1}}} = \frac{x^{1+\frac{1}{x}}}{(x+1)^{1+\frac{1}{x+1}}},$$

and x being indefinitely increased the limiting value is unity.

Now applying the first criterion of the system (A), we have

$$\frac{l\dfrac{1}{u_x}}{lx} = \frac{\dfrac{x+1}{x}lx}{lx} = \frac{x+1}{x},$$

and the limiting value is again unity. Applying the second criterion in (A), we have

$$\frac{l\dfrac{1}{xu_x}}{llx} = \frac{lx^{\frac{1}{x}}}{llx} = \frac{lx}{xllx},$$

the limiting value of which found in the usual way is 0. Hence the series is divergent.

Ex. 6. Resuming the hypergeometrical series of Ex. 4, viz.

$$1+\frac{a}{b}t+\frac{a(a+1)}{b(b+1)}t^2+\frac{a(a+1)(a+2)}{b(b+1)(b+2)}t^3+\ldots \qquad (14),$$

we have in the case of failure when $t=1$,

$$u_x=\frac{a(a+1)\ldots(a+x-1)}{b(b+1)\ldots(b+x-1)}.$$

Therefore
$$\frac{u_{x+1}}{u_x}=\frac{a+x}{b+x},$$

and applying the first criterion of (C),

$$x\left(\frac{u_x}{u_{x+1}}-1\right)=x\left(\frac{b+x}{a+x}-1\right)$$
$$=\frac{(b-a)x}{a+x},$$

which tends to the limit $b-a$. The series is therefore convergent or divergent according as $b-a$ is greater or less than unity.

If $b-a$ is equal to unity, we have, by the second criterion of (C),

$$lx\left\{x\left(\frac{u_x}{u_{x+1}}-1\right)-1\right\}=lx\left\{\frac{(b-a)x}{a+x}-1\right\}$$
$$=\frac{-alx}{a+x},$$

since $b-a=1$. The limiting value is 0, so that the series is still divergent.

It appears, therefore, 1st, that the series (14) is convergent or divergent according as t is less or greater than 1; 2ndly, that if $t=1$ the series is convergent if $b-a>1$, divergent if $b-a\leqq 1$.

It is by no means necessary to resort to the criteria of system (C) in this case. From (13) page 94 we learn that Γx bears a finite ratio to $\sqrt{x}\left(\frac{x}{\epsilon}\right)^x$, and by writing the nth term in the form $\frac{\Gamma b\,\Gamma(a+n)}{\Gamma a\,\Gamma(b+n)}t^n$, it will be found to be comparable with $\frac{t^n}{n^{b-a}}$, whence follows the result found above.

13. We will now examine the series given us by the methods of Chap. V.

By (22) page 100 we have

$$\Sigma\frac{1}{x^2} = C - \frac{1}{x} - \frac{1}{2x^2} - \frac{2}{x^3}\cdot\frac{B_1}{\lfloor 2} + \frac{2.3.4}{x^5}\cdot\frac{B_3}{\lfloor 4} - \dots,$$

$$= C - \frac{1}{x} - \frac{1}{2x^2} - \frac{B_1}{x^3} + \frac{B_3}{x^5} - \dots.$$

Here numerically $\frac{u_{n+1}}{u_n} = \frac{B_{2n+1}}{x^2 B_{2n-1}} = \frac{n^2}{x^2\pi^2}$

ultimately {see (9) page 109},

and thus the series ultimately diverges faster than any diverging geometrical series however large x may be.

As it stands then our results are utterly worthless since we have obtained divergent series as arithmetical equivalents of finite quantities and in order to enable us to approximate to the numerical values of the latter. We shall therefore recommence the investigations of Chap. V, finding expressions for the remainder after any term of the expansion obtained, so that there will always be arithmetical equality between the two sides of the identity, and we shall be able to learn the degree of approximation obtained by examining the magnitude of the remainder or complementary term.

14. The solution of the problem of the convergency or divergency of series that has been given is so complete that it is scarcely possible to imagine how a case of failure could arise. But we have not only obtained a test for con-

vergence, we have also *classified* it. Let us consider for a moment any infinite series. Its nth term u_n must vanish, if the series is convergent, but it must not become a zero of too low an order; otherwise the series will be divergent in spite of u_n becoming ultimately zero. Thus the zero $\frac{1}{n}$ is of too low an order, since $u_n \equiv \frac{1}{n}$ gives a divergent series; $\frac{1}{n^2}$ is of a sufficiently high order, since $u_n \equiv \frac{1}{n^2}$ represents a convergent series. Now the series on page 128 give us a classified list of *forms of zero*. The zeros of any one form are separated by the value $m=1$ into those that are of too low an order for convergency and those that are not. But between any zero value that gives convergency and that corresponding to $m=1$ (which gives divergency) come *all the subsequent forms of zero*. Series comparable with the series produced by giving m any value >1 in the rth class converge infinitely more slowly than those with a greater value of m, but infinitely faster than any similarly related to the $(r+1)^{\text{th}}$ or subsequent classes, whatever value be given to m in the second case. Thus we may refer the convergency of any series to a definite standard by naming the class and the value of m of a series with which it is ultimately comparable.

15. Tchebechef in a remarkable paper (*Liouville*, XVII. 366) has shewn that if we take the prime numbers 2, 3, 5... only, the series

$$F(2)+F(3)+F(5)+\dots$$

will be convergent if the series

$$\frac{F(2)}{\log 2}+\frac{F(3)}{\log 3}+\frac{F(4)}{\log 4}+\dots$$

is convergent. Compare Ex. 10 at the end of the Chapter.

A method of testing convergence is given by Kummer (*Crelle*, XIII.), inferior, of course, to those of Bertrand, &c., but worthy of notice, as it is closely analogous to his method of approximating to the value of very slowly converging series (Bertrand, *Diff. Cal.* 261). It is by finding a function v_n such that $v_n u_n=0$ ultimately, but $\frac{v_n u_n}{u_{n+1}} - v_{n+1} > 0$ when n is ∞. His further paper is in *Crelle*, XVI. 208.

We shall not touch the question of the meaning of divergent series; De Morgan has considered it in his *Differential Calculus*, or an article by Prehn (*Crelle*, XLI. 1) may be referred to.

EXERCISES.

1. Find by an application of the fundamental proposition two limits of the value of the series

$$\frac{1}{a^2+1}+\frac{1}{a^2+4}+\frac{1}{a^2+9}+\dots.$$

In particular shew that if $a=1$ the numerical value of the series will lie between the limits $\frac{\pi}{2}$ and $\frac{\pi}{4}$.

2. The sum of the series

$$\frac{1}{1^{1+\delta}}+\frac{1}{2^{1+\delta}}+\dots$$

(where δ is positive) lies between

$$\frac{2^\delta-\frac{1}{2}}{2^\delta-1} \text{ and } \frac{2^\delta}{2^\delta-1}.$$

3. Examine the convergency of the following series

$$1+e^{-1}+e^{-(1+\frac{1}{2})}+e^{-(1+\frac{1}{2}+\frac{1}{3})}+\dots,$$

$$1+e^{-1}+\frac{1}{2}e^{-(1+\frac{1}{2})}+\frac{1}{3}e^{-(1+\frac{1}{2}+\frac{1}{3})}+\dots,$$

$$e^{u_x}+e^{u_{x+1}}+\dots,$$

$$\frac{\sin x}{1^x}+\frac{\sin\frac{1}{2}x}{2^x}+\frac{\sin\frac{1}{3}x}{3^x}+\dots,$$

$$1+2^{-\frac{2}{3}}+3^{-\frac{3}{4}}+\dots,$$

$$1+2^{-\frac{3}{2}}+3^{-\frac{4}{3}}+\dots,$$

$$1+\frac{1^n}{2^{n+a}}+\frac{2^n}{3^{n+a}}+\dots,$$

$$\left(\frac{\log 2}{1}\right)^a+\left(\frac{\log 3}{2}\right)^a+\dots.$$

4. Are the following series convergent?

$$\frac{3}{2}x+\frac{5}{5}x^2+\frac{7}{10}x^3+\frac{9}{17}x^4+\dots+\frac{2n+1}{n^2+1}x^n+\dots \text{ where } x \text{ is real.}$$

$$1+x\cos\alpha+x^2\cos 2\alpha+\dots \quad \text{where } x \text{ is real or imaginary.}$$

5. The hypergeometrical series

$$1+\frac{ab}{cd}x+\frac{a(a+1)b(b+1)}{c(c+1)d(d+1)}x^2+\dots$$

is convergent if $x<1$, divergent if $x>1$.

If $x=1$ it is convergent only when $c+d-a-b>1$.

6. For what values of x is the following series convergent?

$$x+2^2\frac{x^2}{1.2}+3^3\frac{x^3}{1.2.3}+\dots.$$

7. In what cases is

$$\frac{x^2+x}{x^2+1}\cdot\frac{x^4+x}{x^4+1}\cdot\frac{x^6+x}{x^6+1}\cdot\dots \quad \text{finite?}$$

8. Shew that

$$\frac{1}{u_0}+\frac{1}{u_1}+\frac{1}{u_2}+\dots$$

is convergent if $u_{n+2}-2u_{n+1}+u_n$ be constant or increase with n.

9. If

$$\frac{u_{n+1}}{u_n}=\alpha-\frac{\beta}{n}+\frac{\gamma}{n^2}+\dots,$$

shew that the series converges only when $\alpha<1$, or when $\alpha=1$, and $\beta>1$.

10. A series of numbers $p_1, p_2 \dots$ are formed by the formula

$$n=\frac{p_n}{A\log p_n+B},$$

shew that the series $F(p_1)+F(p_2)+\dots$, will be convergent if $\frac{F(2)}{\log 2}+\frac{F(3)}{\log 3}+\dots$ is convergent.

[Bonnet, *Liouville*, VIII. 73.]

11. Shew that the series

$$a_0 + a_1 + a_2 + \dots,$$

and $\dfrac{a_1}{a_0} + \dfrac{a_2}{a_0 + a_1} + \dfrac{a_3}{a_0 + a_1 + a_2} + \dots\dots\dots$ converge and diverge together.

Hence shew that there can be no test-function $\phi(n)$ such that a series converges or diverges according as $\phi(n) \div u_n$ does not or does vanish when n is infinite.

[Abel, *Crelle*, III. 79.]

12. Shew that if $f(x)$ be such that

$$\frac{xf'(x)}{f(x)} = 1,$$

when $x = 0$, the series $u_1 + u_2 + \dots$ and $f(u_1) + f(u_2) + \dots$ converge and diverge together.

13. Prove from the fundamental proposition Art. 6 that the two series

$$\left.\begin{array}{l}\phi(1) + \phi(2) + \phi(3) + \dots\dots\dots\dots \\ \phi(1) + m\phi(m) + m^2\phi(m^2) + \dots\dots\end{array}\right\}$$ m being positive are convergent or divergent together.

14. Deduce Bertrand's criteria for convergence from the theorem in the last example.

[Paucker, *Crelle*, XLIII. 138.]

15. If $a_0 + a_1x + a_2x^2 + \dots$ be a series in which $a_0, a_1, \dots,$ do not contain x and it is convergent for $x = \delta$ shew that it is convergent for $x < \delta$ even when all the coefficients are taken with the positive sign.

16. The differential coefficient of a convergent series remains finite *within* the limits of its convergency. Examine the case of $u_n \equiv \phi(n) \cos n\theta$. Ex. $\phi(n) \equiv \dfrac{1}{n}$, when the sum of the original series is $-\dfrac{1}{2}\log(2 - 2\cos x)$.

17. Find the condition that the product $u_1 u_2 u_3 \ldots$ should be finite.

Ex. $2^{\frac{1}{2}} . 3^{\frac{1}{3}} . 4^{\frac{1}{4}} \ldots$

18. If the series $u_0 + u_1 + u_2 + \ldots$ has all its terms of the same sign and converges, shew that the product

$$(1 + u_0)(1 + u_1) \ldots \text{ is finite.}$$

Shew that this is also the case when the terms have not all the same sign provided the series and that formed by squaring each term both converge.

[Arndt, *Grunert*, XXI. 78.]

CHAPTER VIII.

EXACT THEOREMS.

1. In the preceding chapters and more especially in Chapter II. we have obtained theorems by expanding functions of Δ, E and D by well-known methods such as the Binomial and Exponential Theorem, the validity of which in the case of algebraical quantities has been demonstrated elsewhere. But this proceeding is open to two objections. In the first place the series is only equivalent to the unexpanded function when it is taken in its entirety, and that is only possible when the series is convergent; so that there can in this case alone be any arithmetical equality between the two sides of the identity given by the theorem. It is true that the laws of convergency for such series when containing algebraical quantities have been investigated, but it is manifestly impossible to assume that the results will hold when the symbols contained therein represent operations, as in the present case. And secondly, we shall very often need to use the method of Finite Differences for the purpose of shortening numerical calculation, and here the mere knowledge that the series obtained are convergent will not suffice; we must also know the degree of approximation.

To render our results trustworthy and useful we must find the limits of the error produced by taking a given number of the terms of the expansion instead of calculating the exact value of the function that gave rise thereto. This we shall do precisely as it is done in Differential Calculus. We shall find the remainder after n terms have been taken, and then seek for limits between which that remainder must lie. We shall consider two cases only—that of the series on page 13 (usually called the *Generalized form of Taylor's Theorem*) and that on page 90. The first will serve for a type of most of the theorems of Chapter II. and deserves notice on account of the relation in which it stands to the fundamental theorem of the Differential Calculus; the close analogy between them will be rendered still more striking by the result of the investigation into the value of the remainder. But it is in the second of the two theorems chosen that we see best the importance of such investigations as these. Constantly used to obtain numerical approximations, and generally leading to divergent series, its results would be wholly valueless were it not for the information that the known form of the remainder gives us of the size of the error caused by taking a portion of the series for the whole.

Remainder in the Generalized form of Taylor's Theorem.

2. Let v_x be a function defined by the identity

$$(x-a)\,v_x \equiv u_x - u_a \qquad (1).$$

By repeated use of the formula

$$\Delta w_w v_x = w_{x+1}\,\Delta v_x + v_x \Delta w_x \qquad (2)$$

we obtain

$$\begin{aligned}(x-a+1)\,\Delta v_x + v_x &= \Delta u_x,\\ (x-a+2)\,\Delta^2 v_x + 2\Delta v_x &= \Delta^2 u_x,\\ \dots\dots\dots\dots &= \dots\dots\\ (x-a+n)\,\Delta^n v_x + n\Delta^{n-1} v_x &= \Delta^n u_x.\end{aligned}$$

Substituting successively for v_x, Δv_x, $\Delta^2 v_x \dots$ we obtain after slight re-arrangement

$$\begin{aligned}u_a = u_x + (a-x)\,\Delta u_x + \frac{(a-x)(a-x-1)}{1.2}\,\Delta^2 u_x + \&\text{c.}\\ + \frac{(a-x)\dots(a-x-n+1)}{\underline{|n}}\,\Delta^n u_x\\ + R_n \qquad (3),\end{aligned}$$

where $$R_n = \frac{(a-x)(a-x-1)\dots(a-x-n)}{\underline{|n}}\,\Delta^n v_x \qquad (4),$$

v_x representing $\dfrac{u_a - u_x}{a-x}$, as is seen from (1).

3. This remainder can be put into many different forms closely analogous, as has been said, to those in the ordinary form of Taylor's Theorem. For instance, if $u_x \equiv f(x)$ we have

$$v_x = \int_0^1 f'\{x + (a-x)z\}\,dz;$$

$$\therefore \Delta^n v_x = \int_0^1 \Delta^n f'\{x + (a-x)z\}\,dz$$

$$= \Delta^n f'\{x + (a-x)\theta\},$$

where θ is some proper fraction.

If we write $x + h$ for a, this last may be written $\Delta^n f'(x + h\theta)$ where Δx is now supposed to be $1-\theta$ instead of unity, and R_n appears under the form

$$\frac{h^{(n+1)}}{\lfloor n}(1-\theta)^n \frac{\Delta^n f'(x+h\theta)}{(\Delta x)^n} \qquad (5),$$

from which we can at once deduce Cauchy's form of the remainder in Taylor's Theorem, i.e.

$$\frac{h^{n+1}}{\lfloor n}(1-\theta)^n f^{n+1}(x+\theta h),$$

after the easy generalization exemplified at the bottom of page 11.

4. Another method of obtaining the remainder is so strikingly analogous to one well known in the Infinitesimal Calculus that we shall give it here. (Compare Todhunter's *Diff. Cal.* 5th Ed. p. 83.)

Let

$$\phi(z) - \phi(x) - (z-x)\Delta\phi(x) - \frac{(z-x)^{(2)}}{\lfloor 2}\Delta^2\phi(x) - \\ - \frac{(z-x)^{(n)}}{\lfloor n}\Delta^n\phi(x)$$

be called $F(x)$; where

$$(z-x)^{(r)} \equiv (z-x)(z-x-1)\dots(z-x-r+1).$$

Then, since from (2)

$$\Delta\left[\frac{(z-x)^{(r)}}{\underline{|r}}\Delta^r\phi(x)\right]=\frac{(z-x-1)^{(r)}}{\underline{|r}}\Delta^{r+1}\phi(x)-\frac{(z-x-1)^{(r-1)}}{\underline{|r-1}}\Delta^r\phi(x),$$

we obtain

$$\Delta F(x)=-\frac{(z-x-1)^{(n)}}{\underline{|n}}\Delta^{n+1}\phi(x) \qquad (6).$$

Now if $z-x$ be an integer

$$F(z)-F(x)=\Delta F(x)+\Delta F(x+1)+\ldots+\Delta F(z-1) \qquad (7),$$

and hence is equal to the product of $(z-x)$ and some quantity intermediate between the greatest and least of these quantities, and as $\Delta F(x)$ is supposed to change continuously through the space under consideration, it will at some point between x and z (we might say between x and $\overline{z-1}$), take the value in question, and we may thus write (7),

$$F(z)-F(x)=(z-x)\,\Delta F\{z+\theta(x-z)\}.$$

But $F(z)=0$, $\therefore$ (6) becomes

$$F(x)=-(z-x)\,\Delta F\{z+\theta(x-z)\}$$

$$=\frac{(z-x)\,\{\theta(z-x)-1\}^{(n)}}{\underline{|n}}\Delta^{n+1}\{z+\theta(x-z)\},$$

or, if $z-x=h$,

$$=\frac{(\theta h)^{(n+1)}}{\theta\,\underline{|n}}\Delta^{n+1}(x+h-\theta h) \qquad (8).$$

5. A more useful form of the result would be derived at once by summing both sides of (6), remembering that $F(z)$ is zero. Since $(z-x-1)^{(n)}$ is positive for all values of x less than z, we see that $F(x)$ lies between the products of the sum of the coefficients of the form $\dfrac{(z-x-1)^{(r)}}{\underline{|r}}$ by the greatest and least values of $\Delta^{n+1}\phi(x)$. But the sum in question is $\dfrac{(z-x)^{(n+1)}}{\underline{|n+1}}$, so that the form thus obtained is very convenient.

This last investigation only applies when $z - x$ is an integer, or in other words when the series would terminate. It is evident that if it were not so we could not draw conclusions as to the magnitude of $F(z) - F(x)$ from the successive differences as we do above. The form of the periodical constant would affect $F(z) - F(x)$ without affecting the other side of the equation.

Remainder in the Maclaurin Sum-formula.

6. In finding the remainder in the Maclaurin sum-formula we shall take it in the slightly modified form obtained by writing u_x for $\int u_x dx$ and performing Δ on both sides. It then becomes

$$\frac{du_x}{dx} = \Delta u_x - \frac{1}{2}\Delta\frac{du_x}{dx} + \frac{B_1}{1.2}\Delta\frac{d^2u_x}{dx^2} - \dots \quad (9),$$

but for convenience we shall write it in the more symmetrical form (using accents to denote differentiation)

$$u_x' = \Delta u_x + A_1\Delta u_x' + A_2\Delta u_x'' + \dots + A_{2n-1}\Delta u_x^{2n-1} + R_{2n} \dots (10),$$

where

$$A_1 = -\frac{1}{2},\ A_3 = A_5 = \dots = 0, \text{ and } A_{2r} = (-1)^{r+1}\frac{B_{2r-1}}{\lfloor 2r}.$$

By Taylor's Theorem we have (Todhunter's *Int. Cal.* Ch. IV.)

$$\Delta u_x = u_x' + \frac{1}{1.2}u_x'' + \dots + \frac{1}{\lfloor 2n}u_x^{2n} + \int_0^1 \frac{z^{2n}}{\lfloor 2n}Pdz,$$

$$\Delta u_x' = u_x'' + \dots + \frac{1}{\lfloor 2n-1}u_x^{2n} + \int_0^1 \frac{z^{2n-1}}{\lfloor 2n-1}Pdz,$$

$$\dots = \dots\dots\dots\dots\dots\dots\dots\dots\dots$$

$$\Delta u_x^{2n-1} = u_x^{2n} + \int_0^1 zPdz,$$

where $P \equiv u_{x+1-z}^{2n+1} \equiv \frac{d^{2n+1}}{dx^{2n+1}}u_{x+1-z}$.

Substitute in (10) and the coefficient of u_x^r is

$$\frac{1}{\underline{|r}} + \frac{A_1}{\underline{|r-1}} + \frac{A_2}{\underline{|r-2}} + \dots + A_{r-1} \qquad (11).$$

This must vanish through the identity expressed in (10). Our symbolical work is the demonstration of this.

The coefficient of Pdx under the integral sign is

$$\frac{z^{2n}}{\underline{|2n}} + A_1 \frac{z^{2n-1}}{\underline{|2n-1}} + \dots + A_{2n-1} z \equiv \phi(2n, z) \text{ suppose.}$$

We shall now shew that $\phi(2n, z)$ does not change sign between the limits of the integral, remains positive or negative as m is even or odd, and has but one maximum (or minimum) value in each case. We see from (11) that $\phi(r, z)$ vanishes when $z = 1$, as it also does when $z = 0$.

7. Assume the above to hold good for some value of n, say an even one, so that $\phi(2n, z)$ is positive between 0 and 1, has but one maximum and vanishes at the limits. Add thereto A_{2n} (which is negative) and integrate and we obtain $\phi(2n+1, z)$. Now this vanishes at both limits, and therefore its differential coefficient $\phi(2n, z) + A_{2n}$ must vanish at some point between them. Now this last is negative at each limit and has but one maximum, thus it must vanish twice,—in passing from negative to positive and from positive to negative,—so that $\phi(2n+1, z)$ has only one minimum followed by a maximum between 0 and 1, and thus can vanish but once. Adding A_{2n+1} (which is zero) to it, for the sake of symmetry, and integrating again we obtain $\phi(2n+2, z)$. This vanishes also at both limits, and its differential coefficient is, as we have seen, at first negative and then positive, changing sign but once. Thus $\phi(2n+2, z)$ has but one maximum and remains positive, which was what we sought to prove. Continuing thus, the theorem is proved for all subsequent values of n, if it be true for any particular one; and as it is true for $\phi(2, z)$ or $\frac{z^2 - z}{2}$, it is generally true.

8. Since $\phi(2n, z)$ retains its sign between the limits

$$R_{2n} = -\int_0^1 \phi(2n, z)\, u^{2n+1}_{x+1-z}\, dz = -u^{2n+1}_{x+\theta} \int_0^1 \phi(2n, z)\, dz, \quad 0 < \theta < 1$$

$$= A_{2n} u^{2n+1}_{x+\theta} \text{ in virtue of (11).}$$

Now perform Σ on both sides of (9) and write $\int u_x dx$ for u_x,

$$\Sigma u_x = C + \int u_x dx - \frac{1}{2} u_x + \frac{B_1}{1.2} \frac{du_x}{dx} - \dots + \frac{(-1)^n B_{2n-3}}{\lfloor 2m-2} \frac{d^{2n-3} u_x}{dx^{2n-3}}$$
$$+ \frac{(-1)^{n+1} B_{2n-1}}{\lfloor 2n} \Sigma u^{2n}_{x+\theta}.$$

Let M be the greatest value irrespective of sign that $\frac{d^{2n} u_x}{dx^{2n}}$ has between the limits of summation, x and $x + m$ suppose. Then $\Sigma u^{2n}_{x+\theta}$ must lie between the limits $\pm mM$.

9. Other conclusions may be drawn relative to the size of the error when other facts are known about the behaviour of u_x and its differential coefficients between the limits. For instance, if u_x^{2n} keeps its sign throughout, we may take 0 instead of $-mM$ as one of the limits. The sign of the error will therefore be that of $(-1)^n M$, and, should u_x^{2n+2} keep the same sign as u_x^{2n} between the limits, the error made by taking one term more of the series will have the same sign as $(-1)^{n+1} M$, i.e. the true value will lie between them. This is obviously the case in the series at the top of page 101, hence that series (without any remainder-term) is alternately greater and less than the true value of the function.

10. If u_x^{2n+1} retain its sign between the limits in (10) we have

$$R_{2n} = -\int_0^1 \phi(2n, z)\, u_x^{2n+1} dz = -\phi(2n, \theta)\, \Delta u_x^{2n}, \quad \theta < 1.$$

Now it can be shewn that $\phi(2n, \theta)$ is never greater numerically than $-2A_{2n}$; hence the correction is never so much as twice the next term of the series were it continued instead of being closed by the remainder-term. Thus, wherever we stop, the error is less than the last term, provided that the differential coefficient that appears therein either constantly increases or constantly decreases between the limits taken. This condition is satisfied in all the important series of the form $\Sigma \frac{1}{x^n}$. The series to which they lead on application of the Maclaurin sum-formula all converge for a time and then diverge very rapidly. In spite of this divergence we see that they are admirably adapted to give us approximate values of the sums in question, for we have but to keep the convergent portion and then know that our error is less than the last term we have kept; and by artifices such as that exemplified on page 100, this can be made as small as we like.

11. Several solutions have been given of the problem of finding the remainder after any number of terms of the Maclaurin sum-formula. The one in the text is by Malmstén, and the proof given was suggested by that in a paper by him in *Crelle* (xxxv. 55). It has been chosen because the limits of the error thus obtained are perfectly general and depend on no property of u_x or the differential coefficients thereof, save that such as appear must vary continuously between the limits. The idea of the method used in this very valuable paper was taken from Jacobi, who used it in a paper on the same subject (*Crelle*, XII. 263), entitled *De usu legitimo formulæ summatoriæ Maclaurianæ*. Malmstén's paper contains many other noteworthy results, and in various cases gives narrower limits to the error than those obtained by other processes, while at the same time they are not too complicated. But the whole paper is full of misprints, so that it is better to read an article of Schlömilch (*Zeitschrift*, I. 192), in which he embodies the important part of Malmstén's article, greatly adding to its value by shewing the connection between the remainder and Bernoulli's Function of which we have spoken in Art. 14, page 116. The paper is written with even more than his usual ability, and is to be highly recommended to those who wish further information on the subject.

12. The chief credit of putting the Maclaurin sum-formula on a proper footing, and saving the results it gives from the suspicion under which they must lie as being derived from diverging series, is due to Poisson. In a paper on the numerical calculation of Definite Integrals (*Mémoires de l'Académie*, 1823, page 571) he starts from an expansion by Fourier's Theorem, and obtains for the remainder an expression of the form

$$R_{2n} = -2\left(\frac{-1}{4\pi^2}\right)^n \int_0^x u_z^{2n} \sum_{i=1}^{i=\infty} \frac{1}{i^{2n}} \cos 2i\pi z dz$$

and he then investigates the limits between which this will lie. The investigation is continued by Raabe (*Crelle*, XVIII. 75), and the practical use of the results in the calculation of Definite Integrals examined and estimated, and modifications suitable for the purpose obtained.

A method of obtaining the supplementary term which possesses many advantages is based on the formula

$$\Sigma F(x) = \int F(x)\,dx - \frac{1}{2}F(x) + \frac{1}{\kappa}\int_0^{\infty} \frac{F(x+\kappa z) - F(x-\kappa z)}{e^{2\pi z} - 1}\,dz,$$

where $\kappa \equiv \sqrt{-1}$. On this see a paper by Genocchi (*Tortolini, Ann. Series*, I. Vol. III.), which also contains plentiful references to earlier papers on the subject. Tortolini in the next volume of the same Journal extends it to Σ^n. See also Schlömilch (*Grunert Archiv*, XII. 130).

13. The investigation which appeared in the first edition of this book is subjoined here (Art. 16). The editor thinks that the fundamental assumption, viz. that the remainder may be considered as being equal to

$$\sum_{r=n+1}^{r=\infty} (-1)^{r+1} \frac{B_{2r-1}}{\underline{|2r}} \cdot \frac{d^{2r-1}u_x}{dx^{2r-1}}$$

cannot be held to be legitimate, since the series which the latter represents may be and often is divergent. For the conditions under which the series itself would be convergent, see a paper by Genocchi (*Tortolini, Ann. Series*, I. Vol. VI.) containing references to some results from Cauchy on the same subject. There is a very ingenious proof of the formula itself by integration by parts, in the *Cambridge Mathematical Journal*, by J. W. L. Glaisher, wherein the remainder is found as well as the series, and Schlömilch (*Zeitschrift*, II. 289) has obtained them by a method of great generality, of which he takes this and the Generalized Taylor's Theorem as examples.

14. By far the most important case of summation is that which occurs in the calculation of $\log \Gamma n$ and its differential coefficients. For special examinations of the approximations in this case we may refer to papers by Lipschitz (*Crelle*, LVI. 11), Bauer (*Crelle*, LVII. 256), Raabe (*Crelle*, XXV. 146, and XXVIII. 10). It must be remembered that there is nothing to prevent there being two semi-convergent expansions of the same function of totally different forms, so that the discrepancy noticed by Guderman (*Crelle*, XXIX. 209) in two expansions for $\log \Gamma n$, one of which contains a term in $\frac{1}{n}$, and the other does not, does not justify the conclusion that one must be false.

15. The investigation into the complete form of the Generalized Taylor's Theorem is derived from a paper by Crelle in the twenty-second volume of his Journal. Other papers may be found in *Liouville*, 1845, page 379, (or *Grunert Archiv*, VIII. 166), *Grunert*, XIV. 337, and *Zeitschrift*, II. 269. The convergence and supplementary term of the expansion in inverse factorials (Stirling's Theorem) have also been investigated by Dietrich (*Crelle*, LIX. 163).

The degree of approximation given by transformations of slowly converging series has been arrived at by very elementary work by Poncelet (*Crelle*, XIII. 1), but the results scarcely belong to this chapter.

Limits of the Remainder of the Series for Σu_x. (BOOLE.)

16. Representing, for simplicity, u_x by u, we have

$$\Sigma u = C + \int u dx - \frac{1}{2}u + \frac{B_1}{1.2}\frac{du}{dx} \dots + (-1)^{n-1}\frac{B_{2n-1}}{1.2\dots 2n}\frac{d^{2n-1}u}{dx^{2n-1}}$$

$$+ \Sigma_{r=n+1}^{r=\infty}(-1)^{r-1}\frac{B_{2r-1}}{1.2\dots 2r}\frac{d^{2r-1}u}{dx^{2r-1}}.$$

The second line of this expression we shall represent by R, and endeavour to determine the limits of its value.

Now by (9), page 109,

$$\frac{B_{2r-1}}{1.2\dots 2r} = \frac{2}{(2\pi)^{2r}}\Sigma_{m=1}^{m=\infty}\frac{1}{m^{2r}}.$$

Therefore substituting,

$$R = \Sigma_{r=n+1}^{r=\infty}\Sigma_{m=1}^{m=\infty}\frac{2(-1)^{r-1}}{(2\pi)^{2r}m^{2r}}\frac{d^{2r-1}u}{dx^{2r-1}}$$

$$= 2\Sigma_{m=1}^{m=\infty}\Sigma_{r=n+1}^{r=\infty}\frac{(-1)^{r-1}}{(2m\pi)^{2r}}\frac{d^{2r-1}u}{dx^{2r-1}}.$$

Assume

$$\Sigma_{r=n+1}^{r=\infty}\frac{(-1)^{r-1}}{(2m\pi)^{2r}}\frac{d^{2r-1}u}{dx^{2r-1}} = t.$$

And then, making $\frac{1}{2m\pi} = \epsilon^{\theta}$, we are led by the general theorem for the summation of series (*Diff. Equations*, p. 431) to the differential equation

$$t + \frac{d^2}{dx^2}\epsilon^{2\theta}t = (-1)^n\frac{d^{2n+1}u}{dx^{2n+1}}\epsilon^{(2n+2)\theta},$$

$$\text{or } \frac{d^2t}{dx^2} + (2m\pi)^2 t = \frac{(-1)^n}{(2m\pi)^{2n}}\frac{d^{2n+1}u}{dx^{2n+1}},$$

the complete integral of which is (*Diff. Equations*, p. 383)

$$t = \frac{(-1)^n}{(2m\pi)^{2n+1}}\left\{\sin 2m\pi x\int\cos 2m\pi x\frac{d^{2n+1}u}{dx^{2n+1}}dx\right.$$

$$\left. - \cos 2m\pi x\int\sin 2m\pi x\frac{d^{2n+1}u}{dx^{2n+1}}dx\right\},$$

or, since we have to do only with integer values of x for which $\sin(2m\pi x) = 0$, $\cos(2m\pi x) = 1$,

$$t = \frac{(-1)^{n+1}}{(2m\pi)^{2n+1}}\int\sin 2m\pi x\frac{d^{2n+1}u}{dx^{2n+1}}dx.$$

Hence

$$R = 2\Sigma_{m=1}^{m=\infty}\frac{(-1)^{n+1}}{(2m\pi)^{2n+1}}\int\sin 2m\pi x\frac{d^{2n+1}u}{dx^{2n+1}}dx$$

$$= 2(-1)^{n+1}\int\left\{\frac{\sin 2\pi x}{(2\pi)^{2n+1}} + \frac{\sin 4\pi x}{(4\pi)^{2n+1}} + \dots\right\}\frac{d^{2n+1}u}{dx^{2n+1}}dx \qquad (1),$$

the lower limit of integration being such a value of x as makes $\frac{d^{2n+1}u}{dx^{2n+1}}$ to vanish, the upper limit x. Hence if *within* the limits of integration $\frac{d^{2n+1}u}{dx^{2n+1}}$ retain a constant sign, the value of R will be *numerically* less than that of the function

$$2\int\left\{\frac{1}{(2\pi)^{2n+1}}+\frac{1}{(4\pi)^{2n+1}}\cdots\right\}\frac{d^{2n+1}u}{dx^{2n+1}}dx\,;$$

therefore, than that of the function

$$2\left\{\frac{1}{(2\pi)^{2n+1}}+\frac{1}{(4\pi)^{2n+1}}\ \ldots\ ad\ inf.\right\}\frac{d^{2n}u}{dx^{2n}}\,;$$

therefore, by (9), page 109, than that of the function

$$\frac{1}{2\pi}\frac{B_{2n-1}}{1\,.\,2\,\ldots\,2n}\frac{d^{2n}u}{dx^{2n}}.$$

When n is large this expression tends to a strict interpolation of *form* between the last term of the series given and the first term of its remainder, viz., omitting signs, between

$$\frac{B_{2n-1}}{1\,.\,2\ldots\,2n}\frac{d^{2n-1}u}{dx^{2n-1}}\ \text{and}\ \frac{B_{2n+1}}{1\,.\,2\ldots(2n+2)}\frac{d^{2n+1}u}{dx^{2n+1}} \qquad (2),$$

it being remembered that by (9), page 109, the coefficient of $\frac{d^{2n}u}{dx^{2n}}$ in (1) is, in the *limit*, a mean proportional between the coefficients of $\frac{d^{2n-1}u}{dx^{2n-1}}$ and $\frac{d^{2n+1}u}{dx^{2n+1}}$ in (2). And this interpolation of form is usually accompanied by interpolation of value, though without specifying the form of the function u we can never affirm that such will be the case.

The practical conclusion is that the summation of the convergent terms of the series for Σu affords a sufficient approximation, except when the first differential coefficient in the remainder changes sign within the limits of integration.

DIFFERENCE- AND FUNCTIONAL EQUATIONS.

CHAPTER IX.

DIFFERENCE-EQUATIONS OF THE FIRST ORDER.

1. An ordinary difference-equation is an expressed relation between an independent variable x, a dependent variable u_x, and any successive differences of u_x, as Δu_x, $\Delta^2 u_x \ldots \Delta^n u_x$. The order of the equation is determined by the order of its highest difference; its degree by the index of the power in which that highest difference is involved, supposing the equation rational and integral in form. Difference-equations may also be presented in a form involving successive values, instead of successive differences, of the dependent variable; for $\Delta^n u_x$ can be expressed in terms of u_x, $u_{x+1} \ldots u_{x+n}$.

Difference-equations are said to be linear when they are of the first degree with respect to u_x, Δu_x, $\Delta^2 u_x$, ...; or, supposing successive values of the independent variable to be employed instead of successive differences, when they are of the first degree with respect to u_x, u_{x+1}, u_{x+2}, ... The equivalence of the two statements is obvious.

Genesis of Difference-Equations.

2. The genesis of difference-equations is analogous to that of differential equations. From a *complete primitive*

$$F(x, u_x, c) = 0. \qquad (1),$$

connecting a dependent variable u_x with an independent variable x and an arbitrary constant c, and from the derived equation

$$\Delta F(x, u_x, c) = 0 \qquad (2),$$

we obtain, by eliminating c, an equation of the form

$$\phi(x, u_x, \Delta u_x) = 0 \qquad (3).$$

Or, if successive values are employed in the place of differences, an equation of the form

$$\psi(x, u_x, u_{x+1}) = 0 \qquad (4).$$

Either of these may be considered as a type of difference-equations of the first order.

In like manner if, from a complete primitive

$$F(x, u_x, c_1, c_2, \ldots c_n) = 0 \qquad (5),$$

and from n successive equations derived from it by successive performances of the operation denoted by Δ or E we eliminate $c_1, c_2, \ldots c_n$, we obtain an equation which will assume the form

$$\phi(x, u_x, \Delta u_x, \ldots \Delta_n u_x) = 0 \qquad (6),$$

or the form

$$\psi(x, u_x, u_{x+1}, \ldots u_{x+n}) = 0 \qquad (7),$$

according as successive differences or successive values are employed. Either of these forms is typical of difference-equations of the nth order.

Ex. 1. Assuming as complete primitive $u_x = cx + c^2$, we have, on performing Δ,

$$\Delta u_x = c,$$

by which, eliminating c, there results

$$u_x = x\Delta u_x + (\Delta u_x)^2,$$

the corresponding difference-equation of the first order.

Thus too any complete primitive of the form $u_x = cx + f(c)$ will lead to a difference-equation of the form

$$u_x = x\Delta u_x + f(\Delta u_x) \qquad (8).$$

Ex. 2. Assuming as complete primitive

$$u_x = ca^x + c'b^x,$$

we have

$$u_{x+1} = ca^{x+1} + c'b^{x+1},$$

$$u_{x+2} = ca^{x+2} + c'b^{x+2}.$$

Hence

$$u_{x+1} - au_x = c'(b-a)\,b^x,$$

$$u_{x+2} - au_{x+1} = c'(b-a)\,b^{x+1}.$$

Therefore

$$u_{x+2} - au_{x+1} - b\,(u_{x+1} - au_x) = 0,$$

or

$$u_{x+2} - (a+b)\,u_{x+1} + abu_x = 0 \qquad (9).$$

Here two arbitrary constants being contained in the complete primitive, the difference-equation is of the second order.

3. The arbitrary constants in the complete primitive of a difference-equation need not be absolute constants but only periodical functions of x of the kind whose nature has been explained, and whose analytical expression has been determined in Chap. IV. Art. 4. They are constant with reference only to the operation Δ, and as such, are subject only to the condition of resuming the same value for values of x differing by unity; a condition which however reduces them to absolute constants when x admits only of such systems of values, as for instance in cases when it must be integral.

Existence of a Complete Primitive.

4. We shall now prove the converse of the theorem in Art. 2, viz. that a difference-equation of the nth order implies the existence of a relation between the dependent and independent variables involving n arbitrary constants. We shall do so by obtaining it in the form of a series.

Let us take (6) as the more convenient form of the equation, and suppose that on solving for $\Delta^n u_x$ we obtain

$$\Delta^n u_x = f(x, u_x, \Delta u_x \dots \Delta^{n-1} u_x) \qquad (10).$$

Performing Δ we get

$$\Delta^{n+1} u_x = \text{some function of } x, u_x, \Delta u_x \dots \Delta u_x^n,$$

and on substituting for $\Delta^n u_x$ from (10) this will reduce to an equation of the form

$$\Delta^{n+1} u_x = f_1 (x, u_x, \Delta u_x \dots \Delta^{n-1} u_x) \qquad (11).$$

Continuing this process we shall obtain

$$\Delta^{n+r} u_x = f_r (x, u_x, \Delta u_x \dots \Delta^{n-1} u^x) \qquad (12).$$

But

$$u_r = E^{r+n} u_{-n} = (1+\Delta)^{r+n} u_{-n}$$

$$= u_{-n} + (n+r) c_1 + \frac{(n+r)^{(2)}}{\lfloor 2} c_2 + \dots + \frac{(n+r)^{(n-1)}}{\lfloor n-1} c_{n-1}$$

$$+ \frac{(n+r)^{(n)}}{\lfloor n} f(-n, u_{-n} c_1, \dots c_{n-1})$$

$$+ \dots + f_r(-n, u_{-n}, c_1, \dots c_{n-1}) \qquad (13),$$

where $c_1, c_2, \dots c_{n-1}$ are the values of $\Delta u_{-n} \dots \Delta^{n-1} u_{-n}$, and with the value of u_{-n} form n arbitrary constants in terms of which and r the general value of u_r is expressed. Thus (13) constitutes the general primitive sought. It is evident that it satisfies the equation for $\Delta^p u_x$ for all values of p, since it is derived from those equations.

5. Though this is theoretically the solution of (6) it is practically of but little use. On comparing it with the corresponding theorem in Differential Equations, we see that both labour under the disadvantage of giving the solution

in the form of a series the coefficients of which have to be calculated successively, no law being in general discovered which will give them all. And in one point the series in Differences has the advantage, for it consists of a finite number of terms only, while the other is in general an infinite series. On the other hand, the latter is usually convergent (at all events for small values of r, since the $(m+1)^{\text{th}}$ term contains $\frac{r^m}{\underline{|m}}$ as a factor), so that the first portion of the series suffices. But in our case the last part of the series is as important as the preceding part, since there is no reason to think that the differences will get very small and the factor $\frac{(r+n)^{(m)}}{\underline{|m}}$ is never less than unity.

Having shewn that we may always expect a complete primitive with n arbitrary constants as the solution of a difference-equation of the nth order*, we shall take the case of equations of the first order, beginning with those that are also of the first degree.

Linear Equations of the First Order.

6. The typical form of this class of equations is

$$u_{x+1} - A_x u_x = B_x \qquad (14),$$

where A_x and B_x are given functions of x. We shall first consider the case in which the second member is 0.

To integrate the equation

$$u_{x+1} - A_x u_x = 0 \qquad (15),$$

we have

$$u_{x+1} = A_x u_x,$$

whence, the equation being true for all values of x,

$$u_x = A_{x-1}\, u_{x-1},$$

$$u_{x-1} = A_{x-2}\, u_{x-2},$$

$$\cdots\cdots\cdots\cdots\cdots\cdots$$

$$u_{r+1} = A_r u_r.$$

* An important qualification of this statement will be given in the next chapter.

Hence, by successive substitutions,

$$u_{x+1} = A_x A_{x-1} A_{x-2} \dots A_r u_r \qquad (16),$$

r being an assumed initial value of x.

Let C be the arbitrary value of u_x corresponding to $x = r$ (arbitrary because, it being fixed, the succeeding values of u_x, corresponding to $x = r + 1$, $x = r + 2$, ..., are determined in succession by (15), while u_r is itself left undetermined), then (16) gives

$$u_{x+1} = CA_x A_{x-1} \dots A_r,$$

whence

$$u_x = CA_{x-1} A_{x-2} \dots A_r, \qquad (17),$$

and this is the general integral sought*.

7. While, for any particular system of values of x differing by successive unities, C is an arbitrary constant, for the aggregate of all possible systems it is a periodical function of x, whose cycle of change is completed, while x varies continuously *through* unity. Thus, suppose the initial value of x to be 0, then, whatever arbitrary value we assign to u_0, the values of u_1, u_2, u_3, ... are rigorously determined by the equation (15). Here then C, which represents the value of u_0, is an arbitrary constant, and we have

$$u_{x+1} = CA_x A_{x-1} \dots A_0.$$

Suppose however the initial value of x to be $\frac{1}{2}$, and let E be the corresponding value of u_x. Then, whatever arbitrary

* There is another mode of deducing this result, which it may be well to notice.

Let $u_x = \epsilon^t$. Then $u_{x+1} = \epsilon^{t+\Delta t}$, and (15) becomes

$$\epsilon^{t+\Delta t} - A_x \epsilon^t = 0;$$

$$\therefore \epsilon^{\Delta t} - A_x = 0,$$

whence $\Delta t = \log A_x$,

$$t = \Sigma \log A_x + C$$

$$= \log A_{x-1} + \log A_{x-2} + \dots + C$$

$$= \log \Pi (A_{x-1}) + C, \text{ following the notation of (18).}$$

Therefore

$$u_x = \epsilon^{\log \Pi (A_{x-1}) + C} = C_1 \Pi (A_{x-1})$$

as before.

value we assign to E, the system of values of $u_{\frac{3}{2}}, u_{\frac{5}{2}}, \ldots$ will be rigorously determined by (15), and the solution becomes

$$u_{x+1} = EA_x A_{x-1} \ldots A_{\frac{1}{2}}.$$

The given difference-equation establishes however no connexion between C and E. *The aggregate of possible solutions is therefore comprised in* (17), *supposing* C *therein to be an arbitrary periodical function of* x *completing its changes while* x *changes through unity, and therefore becoming a simple arbitrary constant for any system of values of* x *differing by successive unities.*

We may for convenience express (17) in the form

$$u_x = C\Pi\,(A_{x-1}) \qquad (18),$$

where Π is a symbol of operation denoting the indefinite continued product of the successive values which the function of x, which it precedes, assumes, while x successively decreases by unity.

8. Resuming the general equation (14) let us give to u_x the form above determined, only replacing C by a variable parameter C_x, and then, in analogy with the known method of solution for linear differential equations, seek to determine C_x.

We have
$$u_x = C_x \Pi\,(A_{x-1}),$$
$$u_{x+1} = C_{x+1} \Pi\,(A_x),$$

whence (14) becomes

$$C_{x+1}\Pi\,(A_x) - A_x C_x \Pi\,(A_{x-1}) = B_x.$$

But
$$A_x \Pi\,(A_{x-1}) = \Pi\,(A_x),$$

whence
$$(C_{x+1} - C_x)\,\Pi\,(A_x) = B_x,$$

or,
$$(\Delta C_x)\,\Pi\,(A_x) = B_x,$$

whence
$$\Delta C_x = \frac{B_x}{\Pi\,(A_x)},$$

$$C_x = \Sigma \frac{B_x}{\Pi\,(A_x)} + C \qquad (19);$$

$$\therefore u_x = \Pi(A_{x-1}) \left\{ \Sigma \frac{B_x}{\Pi(A_x)} + C \right\} \qquad (20),$$

the general integral sought*.

Ex. 1. Given $u_{x+1} - (x+1) u_x = 1 . 2 \dots (x+1)$.

From the form of the second member it is apparent that x admits of integral values only.

Here $\quad A_x = x + 1, \quad \Pi(A_{x-1}) = x(x-1) \dots 1,$

$$\frac{B_x}{\Pi(A_x)} = 1, \quad \Sigma \frac{B_x}{\Pi(A_x)} = x;$$

$$\therefore u_x = x(x-1) \dots 1 \times (x + C),$$

where C is an arbitrary constant.

* The simplest method of solving the equation

$$u_{x+1} - A_x u_x = B_x$$

is derived from its analogy with the equation

$$\frac{dy}{dx} + Py = Q.$$

In this latter we sought for a factor u which should make the first side a perfect differential, and found that it was given by solution of the equation

$$\frac{d\mu}{dx} = \mu P.$$

In the present case suppose C_x to be the factor which makes the left-hand side a perfect difference, i.e. of the form $v_{x+1} u_{x+1} - v_x u_x$.

Then $\qquad v_{x+1} = C_x$ and $v_x = A_x C_x$.

Thus

$$v_{x+1} = \frac{v_x}{A_x} = \frac{1}{\Pi(A_x)},$$

as above, putting the arbitrary constant equal to unity, since we only want one integrating factor, not the general expression for such.

Multiplying by v_{x+1} we get

$$\Delta(v_x u_x) = \frac{B_x}{\Pi(A_x)},$$

$$\therefore v_x u_x = \Sigma \frac{B_x}{\Pi(A_x)} + C,$$

$$\therefore u_x = \Pi(A_{x-1}) \left\{ \Sigma \frac{B_x}{\Pi(A_x)} + C \right\}.$$

Ex. 2. Given $u_{x+1} - au_x = b$, where a and b are constant.

Here $A_x = a$, and $\Pi(A_x) = a^x$, therefore

$$u_x = a^{x-1}\left\{\Sigma \frac{b}{a^x} + C\right\} = a^{x-1}\left\{b\frac{1-\frac{1}{a^x}}{1-\frac{1}{a}} + C\right\}$$

$$= \frac{b}{1-a} + C_1 a^x,$$

where C_1 is an arbitrary constant.

We may observe, before dismissing the above example, that when $A_x = a$ the complete value of $\Pi(A_x)$ is a^x multiplied by an indeterminate constant. For

$$\Pi(A_x) = A_x A_{x-1} \dots A_r$$

$$= a \, . \, a \, . \, a \dots (x - r + 1) \text{ times},$$

$$= a^{x-r+1} = a^{-r+1} \times a^x.$$

But were this value employed, the indeterminate constant a^{-r+1} would in one term of the general solution (20) disappear by division, and in the other merge into the arbitrary constant C. Actually we made use of the particular value corresponding to $r = 1$, and this is what in most cases it will be convenient to do.

9. We must here make a remark about the solution of linear equations of the first degree, which will be easily apprehended by those who are acquainted with the analogous property of linear differential equations.

The solution of

$$u_{x+1} - A_x u_x = B_x \qquad (21)$$

consists of two parts, one of which contains the arbitrary constant and is the solution of

$$u_{x+1} - A_x u_x = 0 \qquad (22),$$

and the other is a particular solution of the given equation (21). It is evident that these parts may be found separately; the general solution of (22) being taken, any quantity that satisfies (21) may be added for the second part and the result will be the general solution of (21). It will be often found advisable to use this method in solving such equations, and to guess a particular integral instead of formally solving the equation in its more general form (21).

Ex. 3. Given $\Delta u_x + 2u_x = -x - 1$.

Replacing Δu_x by $u_{x+1} - u_x$, we have

$$u_{x+1} + u_x = -(x+1).$$

Here $A_x = -1$, $B_x = -(x+1)$, whence

$$u_x = C(-1)^x - \frac{x}{2} - \frac{1}{4}.$$

Ex. 4.
$$u_{x+1} - au_x = \frac{a^x}{(x+1)^2}.$$

We find

$$u_x = a^{x-1}\left\{\Sigma \frac{1}{(x+1)^2} + C\right\}$$

$$= a^{x-1}\left\{\frac{1}{1^2} + \frac{1}{2^2} \dots + \frac{1}{x^2} + C\right\}.$$

When, as in the above example, the summation denoted by Σ cannot be effected in *finite* terms, it is convenient to employ as above an indeterminate series. In so doing we have supposed the solution to have reference to positive and integral values of x. The more general form would be

$$u_x = a^{x-1}\left\{\frac{1}{x^2} + \frac{1}{(x-1)^2} \dots + \frac{1}{r^2} + C\right\},$$

r being the initial value of x.

Difference-Equations of the first order, but not of the first degree.

10. The theory of difference-equations of the first order but of a degree higher than the first differs much from that of the corresponding class of differential equations, but it throws upon the latter so remarkable a light, that for this end alone

it would be deserving of attentive study. Before however proceeding to the general theory, we shall notice one or two great classes of such equations that admit of solution by other ways. The analogy between these and well-known forms of differential equations is too evident to need special notice.

A. *Clairault's Form.*

$$u_x = x\Delta u_x + f(\Delta u_x).$$

A solution of this is evidently

$$u_x = cx + f(c),$$

which gives $$\Delta u_x = c.$$

Ex. 5. $$u = x\Delta u_x + (\Delta u_x)^2$$

gives $$u = cx + c^2.$$

B. *One variable absent.*

$$f(\Delta u_x, u_x) = 0.$$

Writing $u_{x+1} - u_x$ for Δu_x and solving we obtain

$$u_{x+1} = \psi(u_x) \text{ suppose};$$

$$\therefore u_{x+2} = \psi(u_{x+1}) = \psi^2(u_x),$$

denoting by $\psi^2(x)$ the result of performing ψ on $\psi(x)$.

Continuing we shall have

$$u_{x+n} = \psi^n(u_x), \text{ or if } u_r = a,\ u_{r+n} = \psi^n(a).$$

This may fairly be called a solution of the equation, but its interpretation and expansion may offer greater difficulties than the original equation presented. This subject will be considered under the head of *Functional Equations.*

Ex. 6. $u_{x+1} = 2u_x^2$; $\therefore u_{x+2} = 2(2u_x^2)^2 = 2^3 u_x^4$,

and continuing we obtain

$$u_{x+n} = \frac{1}{2}(2u_x)^{2^n}.$$

C. *Equations homogeneous in u.*

The type of such equations is

$$f\left(\frac{u_{x+1}}{u_x}, x\right) = 0.$$

Solve for $\frac{u_{x+1}}{u_x}$ and we obtain an equation of the form $\frac{u_{x+1}}{u_x} = A_x$, which leads to a linear equation in u_x.

Ex. 7. $$u_{x+1}^2 - 3u_{x+1}u_x + 2u_x^2 = 0 \qquad (23).$$

Solving $$u_{x+1} = 2u_x \text{ or } u_x,$$

hence $$u_x = 2^x C \text{ or } C.$$

We shall examine further on whether these are the only solutions of (23).

Many other difference-equations may be solved by means of relations which connect the successive values of well-known functions, especially of the circular functions.

Ex. 8. $$u_{x+1}u_x - a_x(u_{x+1} - u_x) + 1 = 0.$$

Here we have

$$\frac{1}{a_x} = \frac{u_{x+1} - u_x}{1 + u_{x+1}u_x}.$$

Now the form of the second member suggests the transformation $u_x = \tan v_x$, which gives

$$\frac{1}{a_x} = \frac{\tan v_{x+1} - \tan v_x}{1 + \tan v_{x+1} \tan v_x}$$

$$= \tan(v_{x+1} - v_x)$$

$$= \tan \Delta v_x,$$

whence

$$v_x = C + \Sigma \tan^{-1}\frac{1}{a_x},$$

$$u_x = \tan\left(C + \Sigma \tan^{-1} \frac{1}{a_x}\right).$$

Ex. 9. Given $u_{x+1}u_x + \sqrt{\{(1 - u_{x+1}{}^2)(1 - u_x{}^2)\}} = a_x$.

Let $u_x = \cos v_x$, and we have

$$\begin{aligned} a_x &= \cos v_{x+1} \cos v_x + \sin v_{x+1} \sin v_x \\ &= \cos(v_{x+1} - v_x) = \cos \Delta v_x, \end{aligned}$$

whence finally

$$u_x = \cos(C + \Sigma \cos^{-1} a_x).$$

But such cases are not numerous enough to warrant special notice, and their solution must be left to the ingenuity of the student. We subjoin examples requiring these and similar devices for their solution.

EXERCISES.

1. Find the difference-equations to which the following complete primitives belong.

1st. $u = cx^2 + c^2$. 2nd. $u = \left\{c(-1)^x - \frac{x}{2}\right\}^2 - \frac{x^4}{4}$.

3rd. $u = cx + c'a^x$. 4th. $u = ca^x + c^2$.

5th. $u = c^2 + c\left(\frac{1-a}{1+a}\right)(-a)^x - \frac{a^{2x+1}}{(1+a)^2}$.

Solve the equations

2. $u_{x+1} - pa^{2x}u_x = qa^{x^2}$.

3. $u_{x+1} - au_x = \cos nx$.

4. $u_{x+1}u_x + (x+2)u_{x+1} + xu_x = -2 - 2x - x^2$.

5. $u_{x+1} - u_x \cos ax = \cos a \cos 2a \ldots \cos(x-2)a$.

6. $u_x u_{x+1} + au_x + b = 0$.

7. $u_x u_{x+1} - au_x + b = 0$.

8. $u_{x+1} - \epsilon^{2x-1}u_x = \epsilon^{x^2}$.

9. $u_{x+1}\sin x\theta - u_x \sin(x+1)\theta = \cos(x-1)\theta - \cos(3x+1)\theta.$

10. $u_{x+1} - au_x = (2x+1)a^x.$

11. $u_{x+1} - 2u_x^2 + 1 = 0.$

12. $(x+1)^2(u_{x+1} - au_x) = a^x(x^2+2x).$

13. $(u_{x+1})^2 = 4(u_x)^2\{(u_x)^2+1\}.$

14. $u_{x+1} = m(u_x)^n.$

15. $\Delta^2 u_x = (u_{x+1})^2 - (u_x)^2.$

16. $u_x \Delta u_x = x\Delta^2 u_x + 1.$

17. $u_{x+1}{}^3 - 3a^2x^2u_{x+1}u_x{}^2 + 2a^3x^3u_x{}^3 = 0.$

18. If P_κ be the number of permutations of n letters taken κ together, repetition being allowed, but no three consecutive letters being the same, shew that

$$\Delta P_\kappa = (n^2 - n)\frac{\alpha^\kappa - \beta^\kappa}{\alpha - \beta},$$

where α, β are the roots of the equation

$$x^2 = (n-1)(x+1).$$ [Smith's Prize.]

CHAPTER X.

GENERAL THEORY OF THE SOLUTIONS OF DIFFERENCE- AND DIFFERENTIAL EQUATIONS OF THE FIRST ORDER.

1. WE shall in this Chapter examine into the nature and relations of the various solutions of a Difference-equation of the first order, but not necessarily of the first degree, and then proceed to the solutions of the analogous Differential Equations in the hope of obtaining by this means a clearer insight into the nature and relations of the latter.

Expressing a difference-equation of the first order and n^{th} degree in the form

$$(\Delta u)^n + P_1 (\Delta u)^{n-1} + P_2 (\Delta u)^{n-2} \ldots + P_n = 0 \qquad (1),$$

$P_1 P_2 \ldots P_n$ being functions of the variables x and u, and then by algebraic solution reducing it to the form

$$(\Delta u - p_1)(\Delta u - p_2) \ldots (\Delta u - p_n) = 0 \qquad (2),$$

it is evident that the complete primitive of any one of the component equations,

$$\Delta u - p_1 = 0, \quad \Delta u - p_2 = 0 \ldots \Delta u - p_n = 0 \qquad (3),$$

will be a complete primitive of the given equation (1) i. e. *a solution involving an arbitrary constant.* And thus far there is complete analogy with differential equations (*Diff. Equations,* Chap. VII. Art. 1). But here a first point of difference arises. The complete primitives of a differential equation of the first order, obtained by resolution of the equation with respect to $\frac{dy}{dx}$ and solution of the component equations, may without loss of generality be replaced by a single complete primitive. (*Ib.* Art. 3.) Referring to the demonstration of this, the reader will see that it depends mainly upon the fact that the differential coefficient with respect to x of any function of V_1, V_2,... V_n, variables supposed dependent on x, will be linear with respect to the differential coefficients of these dependent variables [*Ib.* (16), (17)]. But this property does not remain if the operation Δ is substituted for that of $\frac{d}{dx}$; and therefore the different complete primitives of a difference-equation cannot be replaced by a single complete primitive*. On the contrary, it may be shewn that out of the complete primitives corresponding to the component equations into which the given difference-equation is supposed to be resolvable, an infinite number of other complete primitives may be evolved corresponding, not to particular component equations, but to a *system* of such components succeeding each other according to a determinate law of alternation as the independent variable x passes through its successive values.

Ex. Thus suppose the given equation to be

$$(\Delta u_x)^2 - (a + x)\,\Delta u_x + ax = 0 \qquad (4),$$

which is resolvable into the two equations

$$\Delta u_x - a = 0, \quad \Delta u_x - x = 0 \qquad (5),$$

and suppose it required to obtain a complete primitive which shall satisfy the given equation (4) by satisfying the first of the component equations (5) when x is an even integer, and the second when x is an odd integer.

* This statement must be taken with some qualification. The reason why the primitives in question $V_1 - C_1 = 0$, $V_2 - C_2 = 0$, ..., can be replaced by the single primitive $(V_1 - C)(V_2 - C)... = 0$ is merely that the last equation *exactly expresses the facts stated by all the others* (viz. that some one of the quantities V_1, V_2,... is constant) and *expresses no more than that*. In a precisely similar way the primitives of a difference-equation of the same kind, being represented by $f_1(x, u_x, C_1) = 0$, $f_2(x, u_x, C_2) = 0$, ..., may be equally well represented by $f_1(x, u_x, C) \times f_2(x, u_x, C) \times ... = 0$. But we shall see that the latter equation must be resolved into its component equations before any conclusion is drawn as to the values of Δu_x. It is not *loss* of generality that is to be feared when we combine the separate primitives into a single one, but *gain*. The new equation is the primitive of an equation of a far higher degree (though still of the first order), and though including the original difference-equation is by no means equivalent to it. We shall return to this point (page 184).

The condition that Δu_x shall be equal to a when x is even, and to x when x is odd, is satisfied if we assume

$$\Delta u_x = a\frac{1+(-1)^x}{2} + x\frac{1-(-1)^x}{2}$$

$$= \frac{a+x}{2} + (-1)^x \frac{a-x}{2},$$

the solution of which is

$$u_x = \frac{ax}{2} + \frac{x(x-1)}{4} + (-1)^x\left(\frac{x-a}{4} - \frac{1}{8}\right) + C,$$

and it will be found that this value of u_x satisfies the given equation in the manner prescribed. Moreover, it is a complete primitive*.

2. It will be observed that the same values of Δu_x may recur in any order. Further illustration than is afforded by Ex. 1 is not needed. Indeed, what is of chief importance to be noted is not the method of solution, which might be varied, but the nature of the connexion of the derived complete primitives with the complete primitives of the component equa-

* To extend this method of solution to any proposed equation and to any proposed case, it is only necessary to express Δu_x as a linear function of the particular values which it is intended that it should receive, each such value being multiplied by a coefficient which has the property of becoming equal to unity for the values of x for which that term becomes the equivalent of Δu_x, and to 0 for all other values. The forms of the coefficients may be determined by the following well-known proposition in the Theory of Equations.

PROP. If $\alpha, \beta, \gamma, \ldots$ be the several n^{th} roots of unity, then, x being an integer, the function $\dfrac{\alpha^x+\beta^x+\gamma^x\ldots}{n}$ is equal to unity if x be equal to n or a multiple of n, and is equal to 0 if x be not a multiple of n.

Hence, if it be required to form such an expression for Δu_x as shall assume the particular values $p_1, p_2, \ldots p_n$ in succession for the values $x=1$, $x=2, \ldots x=n$, and again, for the values $x=n+1$, $x=n+2, \ldots x=2n$, and so on, *ad inf.*, it suffices to assume

$$\Delta u_x = P_{x-1}p_1 + P_{x-2}p_2 \ldots + P_{x-n}p_n \qquad (6),$$

where

$$P_x = \frac{\alpha^x+\beta^x+\gamma^x\ldots}{n},$$

$\alpha, \beta, \gamma, \ldots$ being as above the different n^{th} roots of unity. The equation (6) must then be integrated.

tions into which the given difference-equation is resolvable. It is seen that any one of those derived primitives would geometrically form a sort of connecting envelope of the loci of what may be termed its *component* primitives, i.e. the complete primitives of the component equations of the given difference-equation.

If x be the abscissa, u_x the corresponding ordinate of a point on a plane referred to rectangular axes, then any particular primitive of a difference-equation represents a system of such points, with abscissæ chosen from a definite system differing by units, and a complete primitive represents an infinite number of such systems, the system of abscissæ being the same for all. Now let two consecutive points in any system be said to constitute an *element* of that system, then it is seen that the successive elements of a *derived* primitive (according to the definitions implied above) will be taken in a determinate cyclical order from the elements of systems corresponding to what we have termed its *component* primitives.

3. It is possible also to deduce new *complete* primitives from a single complete primitive, provided that in the latter the expression for u_x be of a higher degree than the first with respect to the arbitrary constant. The method, which consists in treating the constant as a variable parameter, and which leads to results of great interest from their connexion with the theory of Differential Equations, will be exemplified in the following section.

Solutions derived from the Variation of a Constant.

A given complete primitive of a difference-equation of the first order being expressed in the form

$$u = f(x, c) \qquad (7),$$

let c vary, but under the condition that Δu shall admit of the same expression in terms of x and c as if c were a constant. It is evident that if the value of c determined by this condition as a function of x be substituted in the given primitive (7) we shall obtain a new solution of the given equation of differences. The process is *analogous* to that by which from the complete primitive of a differential equation we deduce the singular solution, but it differs as to the character of the result. The solutions at which we arrive are not singular solutions, but new complete primitives, the condition to which c is made subject leading us not, as in the case of differential equations, to an algebraic equation for its discovery, but to a difference-equation, the solution of which introduces a new arbitrary constant.

The new complete primitive is usually termed an indirect integral*.

Ex. The equation $u = x\Delta u + (\Delta u)^2$ has for a complete primitive

$$u = cx + c^2 \qquad (8),$$

an indirect integral is required.

Taking the difference on the hypothesis that c is constant, we have

$$\Delta u = c;$$

and taking the difference of (8) on the hypothesis that c is an unknown function of x, we have

$$\Delta u = c + (x+1)\,\Delta c + 2c\Delta c + (\Delta c)^2.$$

Whence, equating these values of Δu, we have

$$\Delta c\,(x + 1 + 2c + \Delta c) = 0 \qquad (9).$$

Of the two component equations here implied, viz.

$$\Delta c = 0, \quad \Delta c + 2c + x + 1 = 0,$$

the first determines c as an arbitrary constant, and leads back to the given primitive (8); the second gives, on integration,

$$c = C(-1)^x - \frac{x}{2} - \frac{1}{4} \qquad (10),$$

* We shall see reason to doubt the propriety of giving to it any special name that would seem to imply that it stood in a special relation to the original difference-equation.

C being an arbitrary constant, and this value of c substituted in the complete primitive (8) gives on reduction

$$u=\left\{C(-1)^x-\frac{1}{4}\right\}^2-\frac{x^2}{4} \qquad (11).$$

Now this is an indirect integral. We see that the *principle* on which its determination rests is that upon which rests the deduction of the singular solutions of differential equations from their complete primitives. But in *form* the result is itself a complete primitive; and the reader will easily verify that it satisfies the given equation of differences without any particular determination of the constant C.

Again, as by the method of Art. 1 we can deduce from (9) an infinite number of complete primitives determining c, we can, by the substitution of their values in (8), deduce an infinite number of indirect integrals of the equation of differences given.

4. The process by which from a given complete primitive we deduce an indirect integral admits of geometrical interpretation.

For each value of c the complete primitive $u=f(x, c)$ may be understood to represent a system of points situated in a plane and referred to rectangular co-ordinates; the changing of c into $c+\Delta c$ then represents a transition from one such system to another. If such change leave unchanged the values of u and of Δu corresponding to a particular value of x, it indicates that there are two consecutive points, i.e. an *element* (Art. 2) of the system represented by $u=f(x, c)$, the position of which the transition does not affect. And the successive change of c, as a function of x ever satisfying this condition, indicates that each system of points formed in succession has one element common with the system by which it was preceded, and the next element common with the system by which it is followed. The system of points formed of these consecutive common elements is the so-called *indirect integral*, which is thus seen to be a connecting envelope of the different systems of points represented by the given complete primitive.

5. It is proper to observe that indirect integrals may be deduced from the difference-equation (provided that we can effect the requisite integrations) without the prior knowledge of a complete primitive.

Ex. Thus, assuming the difference-equation,

$$u = x\Delta u_x + (\Delta u_x)^2 \qquad (12),$$

and taking the difference of both sides, we have

$$\Delta u_x = \Delta u_x + x\Delta^2 u_x + \Delta^2 u_x + 2\Delta u_x \Delta^2 u_x + (\Delta^2 u_x)^2;$$

$$\therefore\ \Delta^2 u_x (\Delta^2 u_x + 2\Delta u_x + x + 1) = 0,$$

which is resolvable into

$$\Delta^2 u_x = 0 \qquad (13),$$

$$\Delta^2 u_x + 2\Delta u_x + x + 1 = 0 \qquad (14).$$

The former gives, on integrating once,

$$\Delta u_x = c,$$

and leads, on substitution in the given equation, to the complete primitive (8).

The second equation (14) gives, after one integration,

$$\Delta u_x = C(-1)^x - \frac{x}{2} - \frac{1}{4}, \qquad (15),$$

and substituting this in (12) we have on reduction

$$u_x = \left\{C(-1)^x - \frac{1}{4}\right\}^2 - \frac{x^2}{4}, \qquad (16),$$

which agrees with (11).

6. A most important remark must here be made. The method of the preceding article is in *no respect analogous to the derivation of the singular solution from the differential equation.* It is precisely analogous to Lagrange's method of solving differential equations by differentiation (Boole, *Diff. Eq.* Ch. VII. Art. 9), where we form by differentiation a differential equation of the second order, (of which the given equation is one of the first integrals,) obtain by integration the

other first integral, and eliminate $\frac{dy}{dx}$ between them. Thus if we have

$$2x\frac{dy}{dx} - y = 0,$$

we obtain

$$\frac{dy}{dx} + 2x\frac{d^2y}{dx^2} = 0,$$

an integral of which is

$$x\left.\frac{dy}{dx}\right|^2 = c,$$

and hence the solution of the given equation is

$$y^2 = 4cx.$$

As a natural consequence of this analogy all the results of this method are solutions of the original difference-equation. It will be remembered on the contrary that the results of the process of finding singular solutions from the differential equation may not be solutions at all. The analogies of this last process will be referred to later in Art. (21).

7. The second equation (14) might have been integrated in another way, i.e. by simply performing Σ upon it. We should then have obtained

$$\Delta u_x + 2u_x + \frac{x(x+1)}{2} = c \qquad (17).$$

Substituting this in (12) we obtain

$$u_x = \Delta u_x (x + \Delta u_x) = \left(c - 2u_x - \frac{x^2}{2}\right)^2 - \frac{x^2}{4} \qquad (18).$$

This appears to be a third complete integral, but it is only another form of (11), which may be written thus

$$u + \frac{x^2}{4} = C^2(-1)^{2x} - \frac{1}{2}C(-1)^x + \frac{1}{16};$$

$$\therefore C(-1)^x - \frac{1}{4} = 2\left\{-u - \frac{x^2}{4} - \frac{1}{16} + C^2(-1)^{2x}\right\}$$

$$= -2u - \frac{x^2}{2} + c',$$

since $C(-1)^{2x}$ *is constant as far as the operation* Δ *is concerned.*

Substituting in (11) we obtain a result equivalent to (18)*.

General Theory of Difference-Equations of the first order and their solutions.

8. We shall now examine the meaning and relationship of difference-equations, their complete primitives and indirect integrals; and to render our ideas clearer shall notice first the analogous cases in differential equations.

If we have a differential equation of the first order and first degree $\frac{dy}{dx}$ has but one value at each point, and the solution consists of a series of curves one of which passes through every point and *no two cut;* for if two members of the family of curves coincided in one point they would coincide during the remainder of their course. But if $\frac{dy}{dx}$ be given by an equation of a higher (suppose the nth) degree this is not the case. Writing the equation in the form

$$\left(\frac{dy}{dx}-p_1\right)\left(\frac{dy}{dx}-p_2\right)\cdots\left(\frac{dy}{dx}-p_n\right)=0 \qquad (19),$$

we see that at every point $\frac{dy}{dx}$ may have any one of the values $p_1, p_2 \cdots p_n$, but must have one of them.

9. This and only this is told us by (19); the statement at the end of the last paragraph is identically the same as the statement contained in (19). Hence anything further that we can extract from (19) must come from laws independent of

* It may be shewn independently, that if one integral of (14) gives a complete primitive, the other must give the same. For if (17) hold, the solution must come under the complete primitive of (14), involving two arbitrary constants. But for all such solutions, (15) must also hold. Hence all solutions derived from (17) and (12) must come among those derived from (15) and (12), and as the converse proposition is also true, the results of the two methods must be identical. This can only be asserted when (14) is of the first degree in $\Delta^2 u_x$; in all other cases we shall see that there is no single complete primitive.

this special equation, which impose conditions on the systems of values that $\frac{dy}{dx}$ can take. The law that effects this is the law of continuity, which requires that $\frac{dy}{dx}$ should vary *continuously*, or that there should not be a finite change in $\frac{dy}{dx}$ corresponding to indefinitely small changes in x and y. Thus if we would trace out a continuous curve that shall be a solution of the equation, and commence moving in the direction given by $\frac{dy}{dx}=p_1$, we shall be compelled to continue moving in the direction given by $\frac{dy}{dx}=p_1$ at each point, and shall not be able to change to the direction $\frac{dy}{dx}=p_2$ at any point* even though motion in that direction is equally contemplated in equation (19). Thus the law of continuity renders equation (19) the same as the system of equations

$$\frac{dy}{dx}-p_1=0, \quad \frac{dy}{dx}-p_2=0, \ldots, \frac{dy}{dx}-p_n=0 \qquad (20),$$

and permits us to solve them separately and take the combined results as forming the solution of (19).

10. Now take the case of difference-equations. As before, if Δu_x or Δy be given uniquely by the given equation, there exist definite point-systems beginning with any point arbitrarily chosen, but entirely fixed by the choice of it. But when the equation is of the form

$$(\Delta y-p_1)(\Delta y-p_2)\ldots(\Delta y-p_n)=0 \qquad (21),$$

Δy may have any of the n values $p_1, p_2, \ldots, p_n$ at each point. And, as before, this and this only is told us by (21), and any further information must be gained by consideration of the general laws that govern Δy and not from the special case before us.

* This is purposely overstated. A case of exception will be noticed later. Art. 20.

11. But here no law of continuity comes to our aid. The changes in x and y are finite and so will therefore that in Δy generally be. Thus there is no reason why Δy should continue to be equal to p_1 because it is so at the particular point which may be under consideration. In fact, if you will trace out a series of points forming a solution, starting from an arbitrarily chosen point, you have at each point the choice of n different values of Δy, that is, of n different directions in which to go to the next point, and your past choice in no way binds your present*. At most it can be demanded that Δy should be analytically expressible, and that the values should not be arbitrarily chosen at each point, but, as we saw in Art. 1, this merely implies that the succession of values of Δy should obey some law, and places no restriction on what that law shall be. The number of point-systems satisfying the equation is therefore infinite, and must defy all attempts at expression, and the equation (21) reduces to the system of equations

$$\Delta y - p_1 = 0,\ \Delta y - p_2 = 0, \ldots\ \Delta y - p_n = 0 \qquad (22),$$

but we are *not* permitted to solve these separately and take the combined results as the full solution of (21).

12. But in spite of all this, if we integrate separately the various equations contained in (22), the resulting series of n families of point-systems (any one point in the plane forming a part of one member of each system and of only one) has great claims to be called a complete solution of (21). Let it be denoted by

$$f_1(x, y, C_1) = 0,\ f_2(x, y, C_2) = 0, \ldots f_n(x, y, C_n) = 0 \quad (23).$$

In the first place, they together impose exactly the same

* The consideration that the equation

$$(\Delta y - p_1)(\Delta y - p_2)\ldots(\Delta y - p_n) = 0$$

means simply that Δy is at every point equal to one of the quantities $p_1, p_2, \ldots p_n$, gives us the important limitations under which the proof on page 160 of the existence of a complete primitive must be taken. Unless the equation is of the first degree there will at every fresh step be a choice of values for Δu_{n+r}, which will of course affect $\Delta^{n+r}u$, and thus the number of distinct expansions will be infinite. When however we have adopted a law as to the recurrence of the values of Δy, the expansion at once becomes definite.

restraints on the values of Δy that (21) does, since the first member of the series permits it to equal p_1, the second permits it to equal p_2, and so on, and thus if taken as alternative equations they lead to the original equation for Δy. And in the second place, if you stand at any point, the n permissible changes of y will be those of such members of these n point-systems as actually pass through this point. Hence *all permissible elements are elements of members of* (23), and thus all possible solutions of the equation are made up of elements of the point-systems included in (23).

13. That the statements in the last paragraph may be true of any series similar to (23), it is necessary and sufficient that it should at every point give all the admissible values of Δy and no more. But this is attainable in many ways other than by taking the integrals of (22). For instance, if equation (21) be

$$(\Delta y - a)(\Delta y - b) = 0 \qquad (24),$$

it is equivalent to the alternative equations

$$\left.\begin{aligned} \Delta y &= \frac{a+b}{2} + \frac{a-b}{2}(-1)^{x-r} \\ \Delta y &= \frac{a+b}{2} - \frac{a-b}{2}(-1)^{x-r} \end{aligned}\right\}^{*} \qquad (25),$$

where r is some fixed value of x. If then these be integrated, they have exactly the same claim to be considered as constituting a complete solution of (24) as have the solutions of

$$\Delta y - a = 0, \ \Delta y - b = 0 \qquad (26).$$

Thus, following the nomenclature of Art. 2, we see that we shall have sets of n *associated derived primitives*, forming as complete a solution of the equation as the set of n *component primitives*. And in no respect do these solutions yield

* It must not be supposed that the presence of a constant r renders these more or less general than (26). Any expression in finite differences implies that some system of values of x (differing by units) has been chosen, fixing the ordinates on which all our points lie, so that r may be said *to define the space about which we are talking*, and is wholly distinct from a constant that determines y, i.e. the position of the point on some one of those ordinates which form our working-ground.

to the others in closeness of connection with the original equation. Had (24) been given in the form

$$\left\{\Delta y - \frac{a+b}{2} - \frac{a-b}{2}(-1)^{x-r}\right\}\left\{\Delta y - \frac{a+b}{2} + \frac{a-b}{2}(-1)^{x-r}\right\} = 0,$$

as it might equally well have been, the above solutions would have changed places, and the last found would have played the part of component primitives to those obtained from the solution of the factors of (24).

14. But in differential equations the solutions of the differential equations

$$\frac{dy}{dx} - p_1 = 0,\ \frac{dy}{dx} - p_2 = 0, \ldots \frac{dy}{dx} - p_n = 0$$

being supposed to be

$$V_1 - C_1 = 0,\ V_2 - C_2 = 0, \ldots, V_n - C_n = 0 \qquad (27),$$

where C_1, C_2,... C_n are arbitrary constants, the single solution

$$(V_1 - C)(V_2 - C) \ldots (V_n - C) = 0 \qquad (28)$$

can be substituted for them, since the latter signifies that the solution consists of all the curves obtained by giving C all possible values in it. This is obviously tantamount to giving C_1, C_2 ... C_n all possible values in the alternative equations (27) from which (28) is formed, and taking all the curves so given. And this being the case, the differential equation obtained from (28) must be the original differential equation, since (28) comprehends exactly all solutions of it and no more.

15. And the reasoning which permits us to write (28) instead of the system of alternative equations (27), holds when they are solutions of a difference- instead of a differential equation. But it no longer follows that we may use (28) to derive our difference-equation from. This may be seen analytically from the following consideration. Suppose, for simplicity's sake, that V_1, V_2, ... V_n are all linear. The equation obtained by performing Δ on (28) will generally be of the $(n-1)^{\text{th}}$ degree in C and of the nth in Δy. On eliminating C between it and (28), we shall in general obtain an equation of

the n^2 degree in Δy instead of the equation of the nth degree from which we obtained (28). But it may also be seen geometrically thus. Suppose we stand at a point and choose C so that (28) contains the point in virtue of $V_1 - C = 0$ containing it. Then if we put $x+1$ for x in (28) we shall obtain for $y + \Delta y$ all the values of y corresponding to $x+1$ on the curves

$$V_1 - C = 0,\ V_2 - C = 0, \ldots V_n - C = 0,$$

no one of which except the first contains the point at which we start. Take now the value of C which causes $V_2 - C = 0$ to contain the point, we have a similar set of values of Δy, and so on for the rest. All these values will of course be given by the equation for Δy derived from (28) in the ordinary way. Thus we see that in general such an equation as (28) will lead to a difference-equation of a much higher order than the one of which it is a solution, and which permits values of Δy wholly incompatible with that difference-equation. And hence we must in general be content with a system of alternative solutions like (23), or if we combine them as in (28) we must understand that the equation in C must be solved before we can deduce the equation in question. It is by no means necessarily the case that a single equation exists that will lead to the given difference-equation, and even if such a solution exists it does not follow that it is the full solution of the difference-equation.

16. But though it is not necessarily so, it may be so. For instance, the equation $y = cx + c^2$ leads to a difference-equation of the second order, i.e. there are two permissible values of Δy. But substituting in the original equation the co-ordinates of any point, c is found to have two values, so that there are two possible values of Δy corresponding to these two values of c. Hence here the single equation can be taken as a complete substitute for the system of alternative equations with which we are usually obliged to content ourselves. This may fairly be called a complete primitive, but it is by no means the case, as we have seen, that every difference-equation has a complete primitive in this sense of the word. Suppose now two such primitives can be discovered—primitives that it leads to and that lead to it—

then the second one will be what has been named an *indirect integral*. The name is very unfortunate, for regarded as an integral it stands exactly on the same footing as the other complete primitive*.

17. It is obvious that if such integrals exist they must be discoverable by the process of rendering C variable, but assuming that the variation of C will not affect Δy. It must be noticed that any integral of the resulting equation will lead to a new and complete integral of the original equation. We need not wait to get a complete primitive (in the stricter sense of the word) of this equation, a component or derived integral will serve. Nor does the method of deriving them from the difference-equation demand special notice here. We shall see better its meaning and scope by working out fully an example.

18. We have seen that the equation

$$u_x = cx + c^2 \qquad (29)$$

leads to the difference-equation

$$u_x = x\Delta u_x + (\Delta u_x)^2 \qquad (30).$$

Representing, as before, by u_x the ordinate of a point whose abscissa is x, we see that (30) represents a family of point-systems such that at any point there are two values of Δu_x, or, in other words, two points with abscissa $x+1$ that form with the chosen point an *element* of the point-systems (see Art. 2). Now (29) represents also a family of point-systems such that two contain each point, these two having for their distinguishing constants the roots of the equation in c formed from (29), by substituting therein the co-ordinates of the chosen point. Thus (29) and (30) are *co-extensive*, the elements that satisfy (30) are elements of the point-systems included in (29).

* In the first edition of this work an analytical proof was given that, if indirect integrals existed, any one might be taken as the complete primitive, and the others as well as the former complete primitive would appear as indirect integrals. This seems to be unnecessary. Any indirect integral conducts to the difference-equation, i.e. it gives precisely the same liberty of choice for Δy that the complete primitive did. Considering it as the complete primitive, any solutions that satisfy these conditions for Δy are therefore, in relation to it, derived or indirect integrals, according as they do not or do leave to Δy the full liberty that the equation does. From this the proposition is evident.

On solving (30) we obtain

$$\Delta u_x = \sqrt{u_x + \frac{x^2}{4}} - \frac{x}{2}, \text{ or} = -\sqrt{u + \frac{x^2}{4}} - \frac{x}{2} \qquad (31),$$

where $\sqrt{u_x + \frac{x^2}{4}}$ is taken to represent the *numerical value**

* As students are so constantly told that the square root of a quantity has necessarily a double sign, and that it is impossible algebraically to distinguish between them or to exclude one without excluding the other, it is necessary to caution them here that, whatever be the truth of the statement as far as analysis is concerned, it is certainly not true when the functions are represented geometrically, or perhaps we should rather say graphically. Nothing is easier than to distinguish between curves satisfying the equations $y = +\sqrt{c^2 - x^2}$ and $y = -\sqrt{c^2 - x^2}$. It is true that they will not be what we are accustomed to call complete curves, but they will be perfectly definite. And with this understanding it will be evident that the equation $\Delta u_x = +\sqrt{u_x + \frac{x^2}{4}} - \frac{x}{2}$ gives a unique value of Δu_x at every point just as much as if the right-hand side were rational, and it is just as impossible for two members of the family it represents to include the same point without wholly coinciding. But not only does a stipulation such as the one we have made about the sign to be taken with $\sqrt{u + \frac{x^2}{4}}$ remove all indefiniteness geometrically, it also (as must necessarily be the case) removes it arithmetically. As an instance take the theorem in italics.

The next value of

$$+\sqrt{u_x + \frac{x^2}{4}} - \frac{x}{2} \text{ is } + \sqrt{u_x + \Delta u_x + \frac{(x+1)^2}{2}} - \frac{x+1}{2}$$

$$= +\sqrt{u_x + \sqrt{u_x + \frac{x^2}{4}} - \frac{x}{2} + \frac{(x+1)^2}{4}} - \frac{x+1}{2}$$

$$= +\sqrt{u_x + \frac{x^2}{4} + \sqrt{u_x + \frac{x^2}{4}} + \frac{1}{4}} - \frac{x+1}{2}$$

$$= +\sqrt{u_x + \frac{x^2}{4}} + \frac{1}{2} - \frac{x+1}{2} = +\sqrt{u_x + \frac{x^2}{4}} - \frac{x}{2}$$

= its former value.

If at any step the wrong sign had been taken to the square root we should have failed to bring the right result, but by adhering to the stipulation, not only do we obtain the right result, but it forms a rigidly accurate proof of the theorem. It is the neglect of the above principle of the uniqueness of such expressions as $+\sqrt{u + \frac{x^2}{4}} - \frac{x}{2}$ that causes much of the obscurity that surrounds singular solutions in differential equations.

of the square root of $u+\frac{x^2}{4}$. Equation (29) gives us the same values for c. And the result of performing Δ on (29) tells us that $\Delta u_x = c$, in other words *The point-system obtained by taking at each step*

$$\Delta u_x = +\sqrt{u_x+\frac{x^2}{4}}-\frac{x}{2},$$

will keep the latter function wholly unaltered, and thus the solution of this equation is

$$c=\sqrt{u_x+\frac{x^2}{4}}-\frac{x}{2}.$$

In a similar way the solution of

$$\Delta u_x = -\sqrt{u_x+\frac{x^2}{4}}-\frac{x}{2}$$

is
$$c' = -\sqrt{u_x+\frac{x^2}{4}}-\frac{x}{2}.$$

We have divided then our point-systems into two totally distinct families, and elements of members of these families are alone permitted by (30). Now suppose we first choose to take the element given by the first equation of (31), and then we change and take that given by the second. We shall then have

$$\begin{aligned}\Delta u_{x+1} &= -\sqrt{u_{x+1}+\frac{(x+1)^2}{4}}-\frac{x+1}{2}\\ &= -(x+1)-\left\{+\sqrt{u_{x+1}+\frac{(x+1)^2}{4}}-\frac{x+1}{2}\right\}\\ &= -(x+1)-c,\\ \text{or } &= -(x+1)-\Delta u_x, \qquad (32)\end{aligned}$$

since our first element belonged to the family

$$\Delta u_x = +\sqrt{u_x+\frac{x^2}{2}}-\frac{x}{2},$$

or its equivalent

$$c = +\sqrt{u_x + \frac{x^2}{4}} - \frac{x}{2}.$$

Let us for the next element return to the element belonging to the first family. As before,

$$\Delta u_{x+2} = +\sqrt{u_{x+2} + \frac{(x+2)^2}{4}} - \frac{x+2}{2}$$

$$= -(x+2) - \left\{-\sqrt{u_{x+2} + \frac{(x+2)^2}{4}} - \frac{x+2}{2}\right\}$$

$$= -(x+2) - \Delta u_{x+1}, \qquad (33),$$

since the last element was taken from the system

$$\Delta u_x = -\sqrt{u_x + \frac{x^2}{4}} - \frac{x}{2}.$$

(32) and (33) give the same equation, viz.

$$\Delta u_{x+1} = -(x+1) - \Delta u_x, \qquad (34),$$

which is identical with (14), page 177.

This on being integrated leads to the equation

$$\Delta u_x = C(-1)^x - \frac{x}{2} - \frac{1}{4} \qquad (35).$$

The undetermined constant enables us to make it give the right value c for Δu_x at the point chosen, and then Δu_x as given by (35) will continue at each point to have a value permitted by (30), but belonging alternately to each of the two systems of values into which we have divided it. Thus (30) and (35) are both true along the whole of our new solution, and we ought to represent this new solution by them as a system of simultaneous equations. But we know from Algebra that we can take as an equivalent system either of them together with the equation produced, by eliminating Δu_x between them. This last does not involve Δu_x at all and is a complete primitive.

19. It is so obvious that all solutions of a difference-equation must be included in those of the equation obtained by performing Δ on it, that it is natural that we should try to obtain new solutions of

$$(\Delta u_x - p_1)(\Delta u_x - p_2)\dots(\Delta u_x - p_n) = 0 \qquad (36),$$

by this method. The important thing to bear in mind is that which has been illustrated in the foregoing investigation, viz. that all that the method leads to is that Δu_x must either always continue equal to a particular one of the roots p_1, p_2, ... p_n, or it must change so that it jumps from the value of one at one point to the value of another at the next, i.e. $\Delta^2 u_x = \Delta p_r$ or $(p_r)_{x+1} - (p_k)_x$. And it is the alternatives of the latter class that make the sole difference between this method and the method of Lagrange of solving differential equations. In the latter if $\frac{dy}{dx} = p_r$ at a point $d \cdot \frac{dy}{dx}$ can in general only equal dp_r since $\frac{dy}{dx}$ cannot jump from being equal to p_r to being equal to p_k.

20. We say that it can *in general only* equal dp_r. It is only prevented from taking the specified jump by that jump being finite, and hence when we get to a point where $p_k = p_r$ the change is possible. If at the next point p_k is still equal to p_r, $\frac{dy}{dx}$ can change back again to p_k, and so on. This will happen if it should chance that at the point where p_k is equal to p_r the curve $\frac{dy}{dx} = p_k$ is going in the direction of the curve $p_k = p_r$. In this case then there will be a solution analogous to our indirect solutions to difference-equations—its equation will be $p_k = p_r$, and it will only exist when the curves given by $\frac{dy}{dx} = p_r$ touch the curve $p_k = p_r$ at the point where they meet it, or, in other words, if the value of $\frac{dy}{dx}$ derived from $p_k = p_r$ is p_k. Such a solution is termed a singular solution*.

* Few people seem aware of what might be called the rarity of singular solutions. The chances are infinity to one that a differential equation of the first order, but not of the first degree, has no singular solution. As far

21. The question at once suggests itself—are there such singular solutions to difference-equations? But the answer is obvious. If there be any such they are included in the indirect integrals. It is true that they will have a peculiarity. If $p_k = p_r$ gave $\Delta y = p_k$, it is evident that the point-system $p_k = p_r$ might be called a solution of the equation

$$(\Delta y - p_1)(\Delta y - p_2) \dots (\Delta y - p_n) = 0 \qquad (37),$$

in virtue of it satisfying $\Delta y - p_r = 0$ at every point, or of satisfying $\Delta y - p_k = 0$ at every point, or of satisfying them alternately in any cycle. Hence it might with propriety be called a *multiple* solution, since it would appear many times over in the list of solutions. But it can never fail to be included in the complete primitive or its indirect integrals or associated integrals. Poisson (*Journal de l'Ecole Polytechnique*, Tom. VI. p. 60) has written a paper on such solutions. An instance of them is given by the equation

$$\frac{y}{4} = \frac{4^{3x}(\Delta y)^3}{9} - \frac{\Delta y}{3} \qquad (38),$$

of which a complete primitive is

$$y = a\left(\frac{1}{2}\right)^{-2x} - \frac{3}{16}a^3 \qquad (39),$$

and for which he obtains the singular solution

$$y = \pm\frac{8}{9}\left(-\frac{1}{2}\right)^{3x} \qquad (40).$$

If two of the values of Δy given by (38) be equal we must have

$$0 = \frac{(4)^{3x}(\Delta y)^2}{3} - \frac{1}{3};$$

$$\therefore \Delta y = \pm\left(\pm\frac{1}{2}\right)^{3x}.$$

as analysis is concerned it is a mere accident that in certain cases $p_k = p_r$ gives p_r as the value of $\frac{dy}{dx}$. In any equation given for examination, or even in one met with in actual investigations, the chance of the existence of a singular solution is much greater, for it has probably not been written down at random, but has been derived from a complete primitive which represents a family of curves having an envelope.

On substitution in (38) we obtain

$$\frac{y}{4} = \pm\frac{1}{9}\left(\pm\frac{1}{2}\right)^{3x} \pm\frac{1}{3}\left(\pm\frac{1}{2}\right)^{3x} = \pm\frac{2}{9}\left(\pm\frac{1}{2}\right)^{3x};$$

$$\therefore \Delta y = -\frac{7}{9}\left(\frac{1}{2}\right)^{3x}, \text{ or } \pm\left(-\frac{1}{2}\right)^{3x},$$

according as we take the upper or lower sign within the bracket.

Thus $y = \pm\frac{8}{9}\left(-\frac{1}{2}\right)^{3x}$ gives us a singular solution or, as it might better be called, a multiple solution*.

22. Leaving these and returning to the solutions of differential equations, we must remark that not only *may* the change from $\frac{dy}{dx} - p_1 = 0$ to $\frac{dy}{dx} - p_2 = 0$ be made at a point where $p_1 = p_2$ without obtaining a discontinuous curve, but as a rule it actually *is* made in every complete curve that satisfies the equation, provided that a singular solution exists. Take, for instance, the equation $y = cx + c^2$, this leads to the alternative differential equations

$$\left.\begin{aligned} &\frac{dy}{dx} = +\sqrt{y + \frac{x^2}{4}} - \frac{x}{2} \\ \text{and} \quad &\frac{dy}{dx} = -\sqrt{y + \frac{x^2}{4}} - \frac{x}{2} \end{aligned}\right\} \qquad (41),$$

and the singular solution is of course

$$\sqrt{y + \frac{x^2}{4}} = 0, \text{ or } y = -\frac{x^2}{4}.$$

This represents a parabola touching the axis of x at the origin and having its axis in the negative direction of y. The two equations in (41) denote the tangents to it through the chosen point, the first representing the one that makes the algebraically greater angle with the axis of x, since $\frac{dy}{dx}$ is greater along it. Now take a tangent and beginning from $x = -\infty$ move along it. At every point it is the solution of

* As in differential equations the results of this method need not be solutions, but if they are solutions, they are singular solutions. Compare Art. 6.

the second of equations (41), since the other tangent through the point has its $\frac{dy}{dx}$ algebraically greater, as will be seen at once from a figure. But as soon as it has touched the envelope it takes at once the *rôle* of being the solution, at every point of its length, of the *first* of equations (41). So that if we take the complete curve, i.e. the whole of the tangent line, as a solution of the equation, we shall have changed from satisfying the second of the alternative equations to satisfying the first; the change taking place at the point of contact with the envelope.

23. This enables us to see very clearly that the envelope is in reality an indirect integral. For let us start from a point on a tangent just *before* it meets the envelope and proceed along it—of course in the positive direction of x—to a point on it just *after* it meets the envelope. Our path at first satisfied the *second* and now satisfies the *first* of equations (41). Let us now change and take the path through the point at which we now are that satisfies the *second* of those equations. It will be the tangent through the point which is *just* going to touch the envelope. On continuing this process we see that we have a circumscribing polygon, the limit of which when the sides are indefinitely diminished is the curve. And this was generated by pursuing exactly the same method that we observe in obtaining derived or indirect integrals from component integrals or complete primitives, viz. by alternating between different solutions*.

24. It will not be necessary to dwell upon the derivation of

* The *Singular Solution* (or rather *Multiple Integral*) of Art. 21 partakes, as we have seen, of the nature of the singular solution of a Differential Equation, since it is derived from the difference equation in the same way, viz. by taking the condition that two of the alternative solutions should coincide. And hence it is not to be wondered at that the singular solution of a differential equation should have somewhat in it of a multiple integral. In point of fact, portions of it form part of all solutions of the original equation. For instance, in the case we are considering the solution of, $\frac{dy}{dx} = -\sqrt{y + \frac{x^2}{4}} - \frac{x}{2}$ is obtained by always choosing the one of the two permissible paths that lie most to the right, supposing that we start from a point in the third quadrant. This takes you in a straight line as far as the curve and then takes you round during the rest of your motion, since any departure must be along a tangent, i.e. more to the left than along the curve.

indirect integrals or singular solutions from the complete primitive. What has been said will be guide sufficient. But before leaving this part of the subject we will examine how far these views enable us to explain the anomalies connected with Singular Solutions in Differential Equations. Boole (*Diff. Eq.* Ch. VIII.) gives the following four Properties of Singular Solutions:

I. An exact differential equation does not admit of a singular solution.

II. The singular solution of a differential equation of the first order and degree renders its integrating factors infinite.

III. A differential equation may be prepared (even without the knowledge of its integrating factors) so as no longer to admit of a given singular solution of the envelope species.

IV. A singular solution will generally make the value of $\frac{d^2y}{dx^2}$ as deduced from the differential equation assume the ambiguous form $\frac{0}{0}$.

The first of these seems self-contradictory. An envelope has the same value of $\frac{dy}{dx}$ as the enveloped curve at the point of contact. Hence it must satisfy the differential equation of the latter, i.e. the equation that gives $\frac{dy}{dx}$. Now the differential equation to any family of curves whatever, say $F(x, y, c) = 0$, can be given in the form of an *exact* equation. All that is necessary is to solve for c and to differentiate the resulting equation $c = \psi(x, y)$. Thus (I.) seems tantamount to saying that no family of curves can have an envelope. (II.) stands or falls with (I.), but is at least remarkable that an *integrating factor* should have any essential connection with that which is represented by the equation. The integrating factor is simply the reciprocal of the factor by which the equation, when in its exact form, was multiplied to bring it into its present form. It is therefore a purely arbitrary thing, and has nought whatever to do with the nature of the equation or with that which it represents. And (III.) is not less puzzling. For since the geometrical envelope has two consecutive points in common with each member of the family, it would seem probable that it would continue to have that property after any transformation of x and y. But were this the case it would continue to touch them all, and thus to be a singular solution according to our previous remark.

25. It cannot be doubted that these anomalies demand explanation, and if our theory of the nature of a singular solution be the right one it must render them intelligible. And from our theory we see no reason why exact differential equations should be more or less likely to have singular solutions than others. It is true that they are of the first degree, and of course no differential equation that gives a single value of $\frac{dy}{dx}$ at every point can have a singular solution (Art. 8). But there is no reason to expect that an exact equation will give one value and one only of $\frac{dy}{dx}$ at every point; it will usually give the value in terms of quantities such as roots of algebraical functions of the co-ordinates, which will have more than one value, and no attempt is made in such equations to limit the interpretation of these to one of their many values. Yet, although our theory declines to take special notice of exact equations, it still gives us a clue to the interpretation of their peculiarity by pointing out a class of equations which possess the property in question, viz. those that give but one value to $\frac{dy}{dx}$ at each point, and which may be for shortness' sake termed *unique* equations. It must be that by our treatment of *exact* equations we make them to all intents and purposes *unique* equations.

26. Let us take the instance given by Boole,

$$x + y\frac{dy}{dx} = \frac{dy}{dx}\sqrt{x^2 + y^2 - a^2}.$$

On dividing by $\sqrt{x^2 + y^2 - a^2}$ to render it an exact equation, we obtain

$$\frac{x + y\dfrac{dy}{dx}}{\sqrt{x^2 + y^2 - a^2}} - \frac{dy}{dx} = 0.$$

Now it is not fair yet to say that this is not satisfied by the singular solution $x^2 + y^2 = a^2$, for that causes the first term to assume the indeterminate form $\frac{0}{0}$; but as soon as we write it in the form $\frac{d}{dx}\sqrt{x^2 + y^2 - a^2} - \frac{dy}{dx} = 0$, we see that the singular solution has ceased to satisfy it, and hence it must be in this step that we have converted the equation into a unique equation. Writing r for $\sqrt{x^2 + y^2 - a^2}$, it becomes $\frac{dr}{dx} - \frac{dy}{dx} = 0$, the integral of which is $y - r = c$, representing a series of parabolas touching the circle $r = 0$. As y is made to increase from its greatest negative value (c being taken positive) r, which at first would generally be negative, gets smaller numerically, vanishes, and then becomes positive. This confirms our remark that the complete curves which are solutions of the equation require $\sqrt{x^2 + y^2 - a^2}$ to be taken partly with a plus and partly with a minus sign, and thus are partly solutions of $+dr - dy = 0$, and partly of $-dr - dy = 0$, the change occurring at the point of contact with the envelope*. Of course this is allowable in consideration that the sign of r is arbitrary at each point, but it will be seen that this stipulation renders the equation a unique equation just as much as the stipulation that r shall always be taken positive.

27. But a difficulty arises here. Since the stipulation, which, as we see, renders the equation unique, enables us to trace out the whole of each curve, it will enable us to trace out all the solutions of the equation, and thus is it not a complete form of the equation? It is true that at any point when two of the curves intersect we shall pass along one or the other according as we reckon that we have or have not passed the point of contact with the envelope, and thus when we make the

* Should this contact not be real, then, so far as real space is concerned, there will be no change in the equation satisfied at every point, and accordingly there will be at no point an alternative path, and therefore no real portion of the singular solution corresponding thereto.

double supposition we shall, by the aid of the stipulation mentioned in the last paragraph, describe the curves without destroying the uniqueness of the equation. But this is equivalent to taking r of double sign at each point, and it is not to be expected that phenomena of intersection (such as singular solutions essentially are) will be discoverable by analysis which calls a point indifferently r, y, and $-r$, y. Whatever stipulation we make as to the sign of r to render $dr - dy = 0$, a unique equation renders it impossible that two such curves should intersect, i.e. should be satisfied by the same values of r and y, but if we consider it an intersection when the one is satisfied by r, y, and the other by $-r$, y, it is not to be expected that our analysis will be equally lax.

28. Assuming then that the true form of the exact differential equation is $dy \pm dr = 0$, we still have to explain how it is that $r = 0$ fails to satisfy the equation. The equation is no longer unique, but the alternative solutions do not seem to assist us, the change from the one to the other implies a sudden change from $\frac{dr}{dy} = 1$ to $\frac{dr}{dy} = -1$. This difficulty, which is merely a particular case of the one arising from (III.), is of a wholly different nature to the last one. We have now at every point precisely the same liberty of path that we had in the original equation—the same number of alternative directions. But we seem unable to change from one set to the other and thus to have no singular solution. Now the sole restrictions on change arise, as we see, from the law of continuity, so that it is in connection with this that the solution of this difficulty must be found. We shall shew how it is that we have no longer the opportunity of choosing, at the points on the singular solution, along which of two paths we shall go.

29. For simplicity's sake, suppose that the appearance of uniqueness in the exact equation is produced, as in the instance that we have taken, by the presence of a quantity of the form $\sqrt{u}$, where u is a rational integral function of x and y, so that $u = 0$ is the singular solution, since it renders equal the two values of $\frac{dy}{dx}$. This is a very common case, and the treatment will apply to other more complicated cases. Let

x, y be the point of contact of a particular primitive with the singular solution, and $x+dx, y+dy$, a neighbouring point on the same primitive. Then since there is tangency with $u=0$ at x, y, the value of u at $x+dx, y+dy$ must be of the second order (and hence $\sqrt{u}$ is of the first order) in dx and dy. Now take $\sqrt{u}$ and x as new variables, η, x, expressing y in terms of them, and draw the curves represented by the primitives when x and η are considered as Cartesian co-ordinates. The axis of η is now the singular solution, and as we proceed along any primitive we find that in its neighbourhood $\dfrac{d\eta}{dx}$ is *finite*, since η was of the first order along a primitive in the neighbourhood of $\eta=0$. Thus the primitives seem to cut $\eta=0$ at an angle. In fact near $u=0$, du was of the order $\sqrt{dx}$ excepting for small displacements in the direction of $u=0$ at the point. Thus $\dfrac{d\eta}{dx}$ is generally infinite for $\eta=0$, or the distortion produced by the new representation is so great that all curves cutting $\eta=0$ in the original will cut it at right angles now. Only those touching it will cut it at a smaller angle, and those that had a yet closer contact will appear to touch it. And, returning to the original, when we remember that $\dfrac{dr}{dx}$ is of the order $\dfrac{1}{\sqrt{dx}}$ for all directions of displacement but one coinciding with $r=0$, we shall see that a solution of the equation

$$\frac{dr}{dx}-\frac{dy}{dx}=0$$

must have the direction given by $r=0$. So considered, the apparent absurdity of saying that $\dfrac{dr}{dx}-\dfrac{dy}{dx}=0$ is satisfied by $r=0, \dfrac{dy}{dx}=0$, passes away. And the preparation which Poisson gives for getting rid of envelopes can be explained on exactly similar principles; it differs chiefly in this, that he has made a rather more general supposition as to the origin of the alternative values of $\dfrac{dy}{dx}$.

30. We might have expected (IV.). The equation for $\frac{d^2y}{dx^2}$, obtained by differentiating the differential equation after solving for $\frac{dy}{dx}$, must give the value of $\frac{d^2y}{dx^2}$ alike for the particular primitive at the point and for the singular solution. And we should not expect these two values to be obtained by giving alternative values to the functions in $\frac{dy}{dx}$ whose values are not unique, since such functions will naturally have unique values on the singular solution. Thus we should expect that the equation for $\frac{d^2y}{dx^2}$ would give an indeterminate result.

We may remark in conclusion that we ought to expect no such anomalies in the solution of difference-equations, as they all arise from change of independent variable, a thing which cannot occur in Finite Differences excepting in the simple form of change of origin.

The Principle of Continuity.

31. We have seen that the great distinction between the subject-matter of Difference- and Differential Equations is, that the law of Continuity rules in the latter and not in the former case. Hence we cannot expect that the results of the former will always be represented in the latter, and we have already dwelt upon cases in which they are not. It will not do to look on the Differential Calculus as a case of the Difference-Calculus, subject merely to the stipulation that the differences are infinitesimally small—while the latter deals with the ratios of simultaneous increments of the dependent and independent variables, the latter deals with the limits which these ratios approach when the increments are indefinitely small—and unless they approach definite limits the case can never be in the province of the Infinitesimal Calculus, however small the differences be taken. We shall now examine

the conditions under which a point-system will merge into a curve, and apply our results to the case of solutions of a difference-equation.

32. It is a familiar but a partial illustration which presents a curve as the limit to which a polygon tends as its sides are indefinitely increased in number and diminished in length. Let us suppose the differences of the value of the abscissa x for the successive points of the polygon to be constant, the law connecting the ordinates of these points to be expressed by a difference-equation, and the corresponding law of the ordinates of the limiting curve to be expressed by a differential equation.

Now there is a more complete and there is a less complete sense in which a curve may be said to be the limit of a polygon.

In the more complete sense not only does every angular point in the perimeter of the polygon approach in the transition to the limit indefinitely near to the curve, but every side of the polygon tends also indefinitely to *coincidence* with the curve. In virtue of this latter condition the value of $\frac{\Delta y}{\Delta x}$ in the polygon tends as Δx is diminished to that of $\frac{dy}{dx}$ in the curve. It is evident that this condition will be realized if the angles of the polygon in its state of transition are all salient, and tend to π as their limit.

But suppose the angles to be alternately salient and re-entrant, and, while the sides of the polygon are indefinitely diminished, to continue to be such without tending to any limit in which that character of alternation would cease. Here it is evident that while every point in the circumference of the polygon approaches indefinitely to the curve, its linear elements do not tend to coincidence of direction with the curve. Here then the limit to which $\frac{\Delta y}{\Delta x}$ approaches in the polygon is not the same as the value of $\frac{dy}{dx}$ in the curve.

33. If then the solutions of a difference-equation of the first order be represented by geometrical loci, and if, as Δx approaches to 0, these loci tend, some after the first, some after the second, of the above modes to continuous curves; then such of those curves as have resulted from the former process and are limits of their generating polygons in respect of the ultimate *direction* of the linear elements as well as position of their extreme points, will alone represent the solutions of the differential equations into which the difference-equation will have merged. This is the geometrical expression of the principle of continuity.

34. The principle admits also of analytical expression. Assuming h as the indeterminate increment of x, let y_x, y_{x+h}, y_{x+2h} be the ordinates of three consecutive points of the polygon, let ϕ be the angle which the straight line joining the first and second of these points makes with the axis of x, ψ the corresponding angle for the second and third of the points, and let $\psi - \phi$, or θ, be called the angle of contingence of these sides.

Now,

$$\tan\phi = \frac{y_{x+h} - y^x}{h}, \quad \tan\psi = \frac{y_{x+2h} - y_{x+h}}{h},$$

$$\tan\theta = \frac{\dfrac{y_{x+2h} - y_{x+h}}{h} - \dfrac{y_{x+h} - y_x}{h}}{1 + \dfrac{y_{x+h} - y_x}{h} \cdot \dfrac{y_{x+2h} - y_{x+h}}{h}}$$

$$= \frac{\dfrac{y_{x+2h} - 2y_{x+h} + y_x}{h}}{1 + \dfrac{y_{x+h} - y_x}{h}\,\dfrac{y_{x+2h} - y_{x+h}}{h}}.$$

Now, since $h = \Delta x$, we have

$$y_{x+h} - y_x = \Delta y_x,$$

$$y_{x+2h} - 2y_{x+h} + y_x = \Delta^2 y_x,$$

$$y_{x+2h} - y_{x+h} = \Delta y_x + \Delta^2 y_x.$$

Therefore replacing y_x by y,

$$\tan\theta = \frac{\dfrac{\Delta^2 y}{\Delta x}}{1 + \left(\dfrac{\Delta y}{\Delta x}\right)^2 + \dfrac{\Delta y}{\Delta x}\dfrac{\Delta^2 y}{\Delta x}} \qquad \text{(A)}.$$

Now the principle of continuity demands that in order that the solution of a difference-equation of the first order may merge into a solution of the limiting differential equation, the value which it gives to the above expression for $\tan\theta$ should, as Δx approaches to 0, tend to become infinitesimal; since in any continuous curve or continuous portion of a curve $\tan\theta$ is infinitesimal. Again, that the above expression for $\tan\theta$ should become infinitesimal, it is clearly necessary and sufficient that $\dfrac{\Delta^2 y}{\Delta x}$ should become so.

35. The application of this principle is obvious. Supposing that we are in possession of any of the complete primitives of a difference-equation in which Δx is indeterminate, then if, in one of those primitives, the value of Δx being indefinitely diminished, that of $\dfrac{\Delta^2 y}{\Delta x}$ tends, *independently of the value of the arbitrary constant* c, to become infinitesimal also, the complete primitive merges into a complete primitive of the limiting differential equation; but if $\dfrac{\Delta^2 y}{\Delta x}$ tend to become infinitesimal with Δx only for a *particular* value of c, then only the *particular* integral corresponding to that value merges into a solution of the differential equation.

36. We have seen that when a difference-equation of the first order has two complete primitives standing in mutual relation of direct and indirect integrals, each of them represents in geometry a system of envelopes to the loci represented by the other. Now suppose that one of these primitives should, according to the above process, merge into a complete primitive of the limiting differential equation, while the other furnishes only a particular solution; then the latter, not being included in the complete primitive of the differential equation, will be a singular solution, and retain-

ing in the limit its geometrical character, will be a singular solution of the envelope species. Hence, the remarkable conclusion that those singular solutions of differential equations which are of the envelope species, originate from particular primitives of difference-equations; their isolation being due to the circumstance that the particular primitives of the difference-equation, obtained from the same complete primitive or indirect integral by taking other values of the arbitrary constant, not possessing that character which is required by the principle of continuity, are unrepresented in the solutions of the differential equation*.

37. Ex. The differential equation $y = x\frac{dy}{dx} + \left(\frac{dy}{dx}\right)^2$ has for its complete primitive

$$y = cx + c^2 \qquad (42),$$

and for its singular solution, which is of the envelope species,

$$y = \frac{-x^2}{4} \qquad (43).$$

It is required to trace these back to their origin in the solutions of a difference-equation. 1st, Taking the difference of the complete primitive, Δx being indeterminate and c a mere constant, we have

$$\Delta y = c\Delta x.$$

Hence $c = \frac{\Delta y}{\Delta x}$, and substituting in the complete primitive,

* It must be remembered that in all this we take no notice whatever of the peculiarities arising from the *periodicity of the arbitrary constant.* The extent of the periodic variations of this constant are wholly independent of the magnitude of Δx, so that they remain the same however small it be, and thus would prove absolutely fatal to the continuity of the resulting curve were C not taken as an absolute constant. But this is in reality no limitation. For we do not pretend that point-systems can ever *become* continuous curves, but they may form the angular points of a polygon of which the curve is the limiting form. Change cannot be continuous in the difference-calculus so that C might be considered an absolute constant since it is constant with reference to the fundamental operation Δ. It is solely because we wish to embrace also the operation D (implying continuous change) in our investigations that we adopt the fiction of C varying continuously subject to the condition of being a *periodic* constant.

we have

$$y = x\frac{\Delta y}{\Delta x} + \left(\frac{\Delta y}{\Delta x}\right)^2 \qquad (44).$$

This is the difference-equation sought.

Taking the difference of (41), Δx being still indeterminate but c a variable parameter, we have as in Ex. Art. 3,

$$\Delta c + 2c = -(x + \Delta x),$$

a difference-equation for determining c, and by precisely the same method as in Ex. Art. 3, we arrive at the solution

$$y = \left\{c(-1)^{\frac{x}{h}} - \frac{h}{4}\right\}^2 - \frac{x^2}{4}$$

$$= c^2 - \frac{hc(-1)^{\frac{x}{h}}}{2} + \frac{h^2}{16} - \frac{x^2}{4} \qquad (45).$$

It results then that (44) has for complete primitives (42) and (45), h being equal to Δx.

2ndly. To determine $\tan\theta$ for the primitive (42), we have

$$\Delta y = c\Delta x, \quad \Delta^2 y = 0,$$

whence, substituting in (A), we find $\tan\theta = 0$. Thus the complete primitive (42) merges without limitation into a complete primitive of the differential equation.

But employing the complete primitive (45), we have

$$\Delta y = hc(-1)^{\frac{x}{h}} - \frac{2xh + h^2}{4},$$

$$\Delta^2 y = -2hc(-1)^{\frac{x}{h}} - \frac{h^2}{2}.$$

Hence

$$\frac{\Delta^2 y}{\Delta x} = -2c(-1)^{\frac{x}{h}} - \frac{h}{2}.$$

Now this value does not tend to 0 as h tends to 0, unless $c=0$. Making therefore $c=0$, $h=0$, in (45), we have as the limiting value of y

$$y=-\frac{x^2}{4},$$

and this agrees with (43).

Thus, while the complete primitive of the differential equation comes without any limitation of the arbitrary constant from the first complete primitive of the difference-equation, the singular solution of the differential equation is only the limiting form of a particular primitive included under the second of the complete primitives (45) of the difference-equation. Geometrically, that complete primitive represents a system of waving or zigzag lines, each of which perpetually crosses and recrosses some one of the system of parabolas represented by the equation

$$y=c^2+\frac{h^2}{16}-\frac{x^2}{4}.$$

As h tends to 0, those lines deviate to less and less distances on either side from the curves; but only one of these tends to ultimate *coïncidence* with its limiting parabola.

38. As the nomenclature of this chapter is not very simple it may be useful to recapitulate the various kinds of solution that a difference-equation of the first order and n^{th} degree may have:

I. Complete primitive} Indirect integral } solutions involving an arbitrary constant from which the equation can be derived, and which can be derived from it. The two classes of solution are the same in their relation to the equation; any one may be chosen as complete primitive, and the next become indirect integrals. Arts 15, 16.

II. Complete primitive (in the less strict sense of the word)} Component primitive } Derived primitive } solutions which do not give to Δu_x all the freedom it may have, but which still allow it such values only as the difference-equation also permits. All these classes of solutions have the same relation to the equation, they are *derived* or *component* in relation to one another. Sets of n such equations granting to Δu_x all the alternative values permitted by the equation form the only complete solution that most equations have, and if the members of any such set be called *component* primitives, all other solutions can be considered as *derived* primitives. Arts. 11—13.

III. Singular Solution, Multiple Integral: See Art. 21.

EXERCISES.

1. Find a complete primitive of the equation

$$(\Delta u_x - a)(\Delta u_x - b) = 0$$

which shall satisfy the equation $\Delta u_x - a = 0$ only when x is a multiple of 3.

2. The equation

$$y = \frac{\Delta y}{2x+1}\left(x^2 + \frac{\Delta y}{2x+1}\right)$$

is satisfied by the complete primitive $y = cx^2 + c^2$. Shew that another complete primitive

$$y = \left\{a(-1)^x - \frac{x}{2}\right\}^2 - \frac{x^4}{4}$$

may thence be deduced.

3. Shew that a linear difference-equation with rational and integral coefficients admits of only one complete primitive.

4. The equation

$$\left(\frac{\Delta y}{a-1}\right)^2 + a^{2x}\frac{\Delta y}{a-1} - a^{2x}y = 0$$

has $y = ca^x + c^2$ for a complete primitive. Deduce another complete primitive.

5. If $u_x u_{x+1} = \dfrac{m}{x+1}$, shew that

$$u_x = \frac{2.4\ldots\ldots(x-1)}{1.3.5\ldots\ldots x} mC \text{ or } \frac{1.3\ldots\ldots(x-1)}{2.4\ldots\ldots xC}$$

according as x is odd or even.

6. Obtain from the difference-equation $y = x\Delta y + \dfrac{m}{\Delta y}$ the indirect integral

$$y = \frac{2.4\ldots\ldots(x-1)}{1.3\ldots\ldots x-2}\,mC + \frac{1.3\ldots\ldots x}{2.4\ldots\ldots(x-1)\,C} \text{ when } x \text{ is odd,}$$

$$y = \frac{1.3\ldots\ldots(x-1)}{2.4\ldots\ldots(x-2)\,C} + \frac{2.4\ldots\ldots x}{1.3\ldots\ldots(x-1)}\,mC \text{ when } x \text{ is even,}$$

and trace the derivation of the singular solution of the differential equation $y = x\dfrac{dy}{dx} + \dfrac{m}{\frac{dy}{dx}}$ therefrom*.

7. From the difference-equation $u = x\Delta u + (\Delta u)^2$ has been derived the indirect integral

$$u = \left\{C(-1)^x - \frac{1}{4}\right\}^2 - \frac{x^2}{4};$$

shew that, assuming this as complete primitive, the equation $u = cx + c^2$ results as indirect integral.

* Here we need not change Δx, but may keep it unity, and suppose that x, y, m, are all infinite and of the same order, since the equation is homogeneous in x, y, and a constant other than that of integration. Substituting in the usual way $\sqrt{2\pi n}\left(\frac{n}{e}\right)^n$ for $\lfloor n$ we shall obtain

$$\Delta^2 y_{2x} = \frac{mC}{2}\sqrt{\frac{\pi}{x}} - \frac{1}{C\sqrt{\pi x}},$$

and, as the work will have shewn that C must be of the same order as $\dfrac{1}{\sqrt{x}}$ so that the terms of this expression are finite, the condition of continuity becomes

$$\frac{mC}{2}\sqrt{\frac{\pi}{x}} - \frac{1}{C\sqrt{\pi x}} = 0 \text{ or } C = \sqrt{\frac{2}{m\pi}},$$

whence $y_{2x} = 2\sqrt{2mx}$, i.e. the point-system becomes the curve $y^2 = 4mx$.

8. The equation $u_{x+1} = (1 + u_x^{\frac{1}{3}})^3$ is satisfied by

$$u_x = (x + c)^3,$$

deduce thence a cycle of three complete primitives.

9. Form the difference-equation whose solution is the system of alternate equations

$$\left.\begin{array}{r} y - cx + x^2 = 0 \\ cy - x + x^2 = 0 \end{array}\right\},$$

and also form a difference-equation of the first order whose complete solution is *one* of the derived integrals of this equation.

10. Shew that if instead of putting equal arbitrary constants in $(V_1 - c_1)(V_2 - c_2) \ldots\ldots = 0$ we put them alternately positive and negative, but of equal numerical value, the resulting differential will be the same, but the resulting difference-equation will be different.

11. Shew that the solution $y = 0$ of the equation

$$\left(\frac{dy}{dx}\right)^3 - 4xy\frac{dy}{dx} + 8y^2 = 0 \text{ (Boole, } \textit{Diff. Eq.}\text{, Ch. VIII.)}$$

is analogous to the singular solutions of difference-equations spoken of in Art. 21.

CHAPTER XI.

LINEAR DIFFERENCE-EQUATIONS WITH CONSTANT COEFFICIENTS.

1. THE type of the equations of which we shall speak in the present chapter is

$$u_{x+n} + A_1 u_{x+n-1} + \ldots + A_n u_x = X \qquad (1),$$

where A_1, A_2,......A_n are constants and X is a function of the independent variable only. This form will manifestly include the form

$$\Delta^n u_x + A_1 \Delta^{n-1} u_x + \ldots, \quad + A_n u_x = X \qquad (2),$$

and may be symbolically written

$$f(E)\, u_x = X \qquad (3),$$

where $f(E)$ is a rational and integral function of E of the nth degree, with unity as the coefficient of the highest term, and with all its coefficients constant.

2. Now we know from (10) page 18 that $E = \epsilon^D$, so that we might write (3) in the form

$$f(\epsilon^D)\, u_x = X \qquad (4),$$

and consider it a linear differential equation of an infinite degree and solve it by the well-known rules for such equations. The *complementary function* would then have an infinity of terms of the form $C\epsilon^{mx}$ where m would be determined by the equation $f(\epsilon^m) = 0$; and to this we should have to add a *particular integral* obtained either by guess or by special rules depending on the form of X. But we shall not adopt this mode of procedure, and that for two reasons. In the first place we have to face the difficulty of an equation of an infinite degree, or rather of an equation that combines the difficulties of transcendental and algebraical equations; and though we know from experience of Ex. 2, page 79, that these difficulties are more apparent than real, and that the infinitude of roots merely signify that the constants obtained are periodic and not absolute constants, the method still is open to the objection of being unnecessarily complex and intricate. But there is a more important reason for not adopting this method. The problems of Finite Differences are really phenomena of *discontinuous* change, the variables do not vary continuously but by jumps. And a method is open to grave objection that treats the change as a continuous one the results of which are inspected only at certain intervals. At all events such a method should not be resorted to when the direct consideration of the operations properly belonging to the Difference-Calculus suffices to solve our problems.

3. We have seen in Chapter II. that E and Δ like D will combine with constant quantities and with one another as though they were symbols of quantity. And thus $f(E)$ when performed on the sum of two quantities gives the same result as if it were performed on each and the results added. Hence if we take any two solutions of the linear difference-equation

$$f(E)\,u_x = 0 \tag{5}$$

the sum of these solutions will also be a solution.

Also any multiple of a solution is obviously a solution. So that if we can obtain n particular solutions $V_1, V_2, \ldots V_n$, connected together by no linear identical relation, then will

$$u_x = C_1V_1 + C_2V_2 + \cdots + C_nV_n \tag{6}$$

be a solution, and in virtue of containing n arbitrary constants it will be the most general solution*. We shall now proceed to find these particular integrals and shall then have solved equation (5), which is the form which (1) assumes when $X=0$.

4. Let $f(E)=0$ have as roots $m_1, m_2, \ldots, m_n$; E being treated as a symbol of quantity. Then we know that

$$f(E) \equiv (E-m_1)(E-m_2) \ldots (E-m_n) \qquad (7),$$

whether E be a symbol of quantity or of operation, so that we may write (5) thus,

$$(E-m_1)(E-m_2) \ldots (E-m_n) u_x = 0 \qquad (8),$$

where $E-m_n \ldots$ denote successive operations *the order of which is indifferent*. But if we solve the equation $(E-m_n) u_x = 0$ we obtain a particular solution of (8), since the operation $(E-m_1)(E-m_2) \ldots (E-m_{n-1})$ performed on a constant of value zero must of course produce zero. Putting in turn each of the other operational factors last, we obtain other particular integrals, and thus when the roots are all different we shall obtain the n particular integrals $V_1, V_2, \ldots V_n$ (which give us by (6) the general solution) by solving n separate equations of the form

$$(E-m) u_x = 0 \qquad (9).$$

5. But if one of the roots is repeated—say r times—this method fails; for r of the solutions would be in point of fact identical or merely multiples of one another. But if the said root be m_κ and we take the *full solution* of the equation

$$(E-m_\kappa)^r u_x = 0 \qquad (10),$$

(involving, as it will, r arbitrary constants), instead of taking the solution of the corresponding case of (9), we shall have as before the right number of arbitrary constants and therefore the most general solution.

* It must be noticed that in linear equations with constant or rational coefficients, there are no difficulties arising from alternative values of the increments of the dependent variables as in the cases which formed the subject of the last chapter. The value given for all successive differences is strictly unique, so that but one complete primitive exists. See note on page 181.

6. We have thus reduced the problem of solving (5) in all cases to that of solving a number of separate equations of the form

$$(E-m)^r u_x = 0 \qquad (11).$$

But (see note page 73)

$$f(E)\, a^x u_x = a^x f(aE)\, u_x \qquad (12);$$

hence

$$(E-m)^r u_x = m^x (mE-m)^r m^{-x} u_x = m^{x+r} \Delta^r (m^{-x} u_x) = 0 \text{ by } (11);$$

$$\therefore \Delta^r (m^{-x} u_x) = 0, \quad \therefore m^{-x} u_x = C_0 + C_1 x + C_2 x^2 + \dots C_{r-1} x^{r-1}$$

since the rth difference of such a function vanishes; and thus

$$u_x = (C_0 + C_1 x + \dots + C_{r-1} x^{r-1})\, m^x \qquad (13).$$

Thus the general solution of (5) is

$$u_x = \Sigma\, (C_0 + C_1 x + \dots C_{r-1} x^{r-1})\, m^x \qquad (14),$$

where r is the number of times the root m is repeated in the equation $f(E) = 0$.

7. We will illustrate the foregoing by an example. Let the equation be

$$u_{x+3} - 3u_{x+1} - 2u_x = 0, \qquad (15),$$

or

$$(E^3 - 3E - 2)\, u_x = 0.$$

This is the same as

$$(E+1)^2 (E-2)\, u_x = 0,$$

and thus the solution of (15) is

$$u_x = (C_0 + C_1 x)(-1)^x + C_2 2^x \qquad (16).$$

8. A slight difficulty presents itself here—not in the theory of the solution, but in the interpretation of the result. It would seem as if we must content ourselves with results impossible in form whenever the roots of the equation for E

are impossible. This may be avoided thus. Impossible roots occur in pairs so that with any term $Cx^r(\alpha+\beta\sqrt{-1})^x$ in the solution, corresponding to a root $(\alpha+\beta\sqrt{-1})$ repeated at least $(r+1)$ times, there will be a term $C'x^r(\alpha-\beta\sqrt{-1})^x$. Assuming

$$\alpha+\beta\sqrt{-1}=\rho(\cos\theta+\sqrt{-1}\sin\theta),$$

which gives

$$\rho=\sqrt{\alpha^2+\beta^2},\quad \tan\theta=\frac{\beta}{\alpha},$$

the terms become

$$x^r\rho^x\{C(\cos x\theta+\sqrt{-1}\sin x\theta)+C'(\cos x\theta-\sqrt{-1}\sin x\theta)\},$$

or

$$x^r\rho^x\{M\cos x\theta+N\sin x\theta\},$$

where M and N are still arbitrary constants. Thus the part of the solution of $f(E)\,u_x=0$ that corresponds to the pair of impossible roots $\alpha\pm\beta\sqrt{-1}$ repeated r times in $f(E)=0$ is

$$(M_0+M_1x+\ldots+M_{r-1}x^{r-1})\,\rho^x\cos x\theta$$
$$+(N_0+N_1x+\ldots+N_{r-1}x^{r-1})\,\rho^x\sin x\theta,$$

which has, as we see, the right number of constants.

Ex. 1. Let the equation be

$$u_{x+6}+2u_{x+3}+u_x=0 \qquad (17),$$

or

$$(E^3+1)^2u_x=0.$$

The roots of $f(E)=0$ are 1, and $\dfrac{1\pm\sqrt{-3}}{2}$, each repeated twice, the solution is therefore

$$c_0+c_1x+(M_0+M_1x)\cos\frac{\pi x}{3}+(N_0+N_1x)\sin\frac{\pi x}{3} \qquad (18),$$

since $\rho=1$ and $\tan\theta=\sqrt{3}$.

9. We have thus obtained a solution of the most general form possible of the equation $f(E)\,u_x=0$. We shall now

proceed to the more general form of equation which we chose as the subject of this chapter, viz.

$$f(E)\, u_x = X \qquad (19).$$

But our past work stands us here in good stead. For if to any solution of this equation we add a solution of $f(E)\, u_x = 0$, the result of performing $f(E)$ upon their sum will be $X + 0$ or X (see Art. 3). If then to a particular solution of (19) we add the general solution of (5), we shall get a solution of (19) involving n arbitrary constants, and which must therefore be the most general solution of (19) possible.

Our task has therefore reduced itself to finding a particular integral of (19). And our first thought is to try if we cannot obtain it by a device similar to that which gave us the solution of (5)—in other words, deduce it from the solutions of simpler equations. At first sight the method seems wholly to fail. For if we solve $(E - m_n)\, u_x = X$ and obtain the solution X_m, it is no longer a solution to the full equation. On performing $f(E)$ upon it, we obtain

$$(E - m_1)(E - m_2)\dots(E - m_{n-1})\, X \qquad (20),$$

which involves X and its next $n-1$ consecutive values.

Similarly if we find X_r the solution of $(E - m_r)\, u_x = X$, we should obtain, on performing $f(E)$ upon it,

$$(E - m_1)\dots(E - m_{r-1})(E - m_{r+1})\dots(E - m_n)\, X \quad (21).$$

10. But a modification of our former method will still give us an integral. Instead of taking merely the solution of one of the simpler equations, take those of all and combine them by multiplying each by a constant and adding the results. If we perform $f(E)$ on $\mu_1 X_1 + \mu_2 X_2 + \dots + \mu_n X_n$—the roots of $f(E) = 0$ being for the present supposed all different—we shall obtain the quantity

$$\{\mu_1 (E - m_2)\dots(E - m_n) + \mu_2 (E - m_1)(E - m_3)\dots(E - m_n) + \dots \\ + \mu_n (E - m_1)\dots(E - m_{n-1})\}\, X \qquad (22).$$

And if by choosing $\mu_1, \mu_2, \dots \mu_n$ aright we are able to make the coefficients of all the powers of E in (22) vanish and the term independent of E become unity, we shall have a solution of (19) in

$$u_x = \mu_1 X_1 + \mu_2 X_2 + \dots \mu_n X_n \tag{23}.$$

To do this we must have

$$\mu_1 (E - m_2)\dots(E - m_n) + \mu_2 (E - m_1)(E - m_3)\dots(E - m_n) + \dots \equiv 1 \tag{24},$$

when *E is treated as a symbol of quantity.* This proviso enables us to divide with confidence by $f(E)$, and we see that

$$\frac{\mu_1}{E - m_1} + \frac{\mu_2}{E - m_2} + \dots + \frac{\mu_n}{E - m_n} \equiv \frac{1}{f(E)} \tag{25},$$

or in other words $\mu_1, \mu_2, \dots$ *are the numerators of the partial fractions into which* $\frac{1}{f(E)}$ *can be resolved.*

11. Nor will this method fail when a root is repeated. Let a root m_κ be repeated r times, then if we use for $X_\kappa, X_{\kappa+1}, \dots X_{\kappa+r-1}$, the solutions of the equations

$$(E - m_\kappa)\, u_x = X,$$
$$(E - m_\kappa)^2\, u_x = X,$$
$$\dots\dots\dots\dots\dots\dots$$
$$(E - m_\kappa)^r\, u_x = X,$$

we shall have for the corresponding values of μ the numerators of the partial fractions forming $\frac{1}{f(E)}$, whose denominators are

$$(E - m_\kappa),\ (E - m_\kappa)^2,\ \dots, (E - m_\kappa)^r.$$

Thus we have reduced the solution of (19) to that of the equation

$$(E - m)\, u_x^r = X \tag{26},$$

which we can write by (12)

$$m^{x+r} \Delta^r (m^{-x} u_x) = X;$$
$$\therefore m^{-x} u_x = \Sigma^r m^{-x-r} X,$$

or $$u_x = m^x \Sigma^r m^{-x-r} X,$$

and (19) is fully solved.

And a little further consideration shews that this last investigation renders unnecessary that in Arts. 2—5, which suggested it. For in each of the quantities X_1, $X_2, \ldots X_n$ there is a term involving an arbitrary constant, and of the form Cm_1^x, Cm_2^x, ... If we include these in the values of X, ... which we substitute in (23) we get the general solution at once*.

12. Let us examine the results at which we have arrived. From the equation $f(E)\, u = X$ we have deduced

$$u_x = \mu_1 X_1 + \ldots + \mu_n X_n \qquad (27),$$

where X_1, $X_2 \ldots$ are the solutions of $(E - m_1)\, u_x = X$ and kindred equations, and μ_1, $\mu_2 \ldots$ are the coefficients of the partial fractions into which $\dfrac{1}{f(E)}$ is resolved *when E is considered a symbol of quantity.* But it is natural to ask,—Could we not have obtained this at once by symbolical methods, thus:—

$$u_x = \frac{1}{f(E)} X = \left\{ \frac{\mu_1}{E - m_1} + \ldots + \frac{\mu_n}{E - m_n} \right\} X \qquad (28).$$

But, since X_1 is a solution of $(E - m_1)\, u_x = X$,

$$X_1 = \frac{X}{E - m_1} \qquad (29),$$

$$\therefore u_x = \mu_1 X_1 + \mu_2 X_2 + \ldots\ldots \mu_n X_n \qquad (30),$$

agreeing with (27).

* It might seem that we shall get more than sufficient constants by this method when roots are repeated. For $(E-m)^r u_x = x$ will give r constants, and $(E - m)^{r-1} u_x = x$ will give $r - 1$ additional ones, while there should only be r in all. But since all the solutions of the equation $(E - m)^{r-1} u_x = 0$ are solutions of the equation $(E - m)^r u_x = 0$, and all the terms which we are considering come from these last equations, we neither gain nor lose in generality whatever solution of $(E - m)^{r-1} u_x = 0$ we take, provided we take the full solution of $(E - m)^r u_x = 0$ which gives r arbitrary constants.

13. At first sight this method seems justified by the properties of E proved in Art. 9, Ch. II. And there is no doubt that, as far as suggestiveness is concerned, such an application of symbolical methods is all that could be desired. But as it stands it is not rigorous. So long as our operations are *direct* we may place absolute reliance on symbolical methods, for the results of the operations are unique, and hence equality in any sense must mean algebraical equality. But so soon as any of the operations are *indirect*, further investigation is needed. The results of the indirect operations are not, in an algebraical point of view, definite, and we must carefully examine each case in order to discover the conditions of interpretation of the results that there may be algebraical equality. For instance,

$$(E-a)(E-b)\,u_x \equiv (E-b)(E-a)\,u_x \qquad (31),$$

but $(E-a)\dfrac{1}{E-a}u_x$ does not equal $\dfrac{1}{E-a}(E-a)\,u_x$ (32),

since the left-hand side is definite and the right-hand side has an arbitrary constant. And, while the first may be taken as an equivalent of u_x, the latter is only so when we stipulate that the constant in the term Ca^x, resulting from the performance of $\dfrac{1}{E-a}$, shall be taken zero. One difficulty of this kind we met with at the beginning of Chapter IV., and we shall content ourselves with investigating the present one, leaving all future cases to the student's own examination.

14. Take then (28). Since u_x is not considered a definite quantity, but as a representative of all the quantities that satisfy (19), there is no absurdity in representing it as equal to the quantity on the right-hand side of (28) which has n undetermined constants. All we have to ask is, whether on performing $f(E)$ on the right-hand side of (28) we shall obtain X; and, this last being a perfectly definite quantity, while the right-hand side of (28) is indefinite, we might expect that some conditions of interpretation would be necessary in (28) to render the equivalence algebraical. But it is not so. For on performing $f(E)$ on the first term, viz. $\dfrac{\mu_1 X}{E-a}$, the opera-

tion $(E-a)$, which is one of those composing $f(E)$*, is *absorbed in rendering this indefinite term strictly definite,* so that the whole result of performing $f(E)$ on it is strictly definite. Thus the result of performing $f(E)$ on the right-hand side of (28) is a strictly definite quantity, and as under some circumstances it must equal X (which we know from the laws of the symbol E), it must be actually equal to it†.

Ex. 2. $$u_{x+2} - 5u_{x+1} + 6u_x = 5^x;$$

$$\text{or } (E-3)(E-2)\,u_x = 5^x;$$

$$\therefore u_x = \frac{5^x}{(E-3)(E-2)} = \left\{\frac{1}{E-3} - \frac{1}{E-2}\right\} 5^x$$

$$= \frac{1}{2}5^x + C3^x - \frac{1}{3}5^x + C'2^x = \frac{1}{6}5^x + C3^x + C'2^x.$$

15. The above is a general solution of linear difference-equations with constant coefficients. But, as we have seen that the part involving arbitrary constants is readily written down after the algebraical solution of the equation $f(E)=0$, and that any particular integral will serve to complete the

* It must be remembered that these operations being direct it is wholly unimportant in what order we perform them.

† While it is true that $f(E)\left\{\frac{\mu_1}{E-m_1} + \dots\right\} X = X$ whatever X may, it is by no means true that $\left\{\frac{\mu_1}{E-m_1} + \dots\right\} f(E)\,X = X$. The importance of care in this respect if we would avoid loose reasoning may be exemplified by an example. In Linear Differential Equations such a quantity as $\frac{\cos mx}{D+a}$ is often evaluated thus:

$$\frac{\cos mx}{D+a} = \frac{(D-a)\cos mx}{D^2-a^2} = \frac{-m\sin mx - a\cos mx}{D^2-a^2} = \frac{-m\sin mx - a\cos mx}{-m^2-a^2}.$$

The first step with the interpretation afforded by the second is wholly inadmissible:

It should be thus:

$$\frac{\cos mx}{D+a} = (D-a)\left\{\frac{\cos mx}{D^2-a^2}\right\} = (D-a)\frac{\cos mx}{-m^2-a^2} = \frac{-m\sin mx - a\cos mx}{-m^2-a^2}.$$

solution, it is usually better to guess a particular integral, or at all events to obtain it by some special method.

The forms of X for which this can readily be done are three, viz.

(I.) When X is of the form a^x. Since $f(E)\,a^x = f(a)\,.\,a^x$ we obviously have $\dfrac{1}{f(E)}\,a^x = \dfrac{1}{f(a)}\,a^x$.

(II.) When X is a rational and integral function of x. Here we have only to expand $f(E)$ in a series of ascending powers of Δ, and perform it in this shape on X. The result* will of course terminate, since X is rational and integral. Should $f(E)$ when expressed in terms of Δ assume the form $\Delta^r(A + B\Delta + \ldots)$, we must evaluate $\dfrac{X}{\Delta^r}$ or $\Sigma^r X$ before applying this method, or may omit the factor Δ^{-r}, apply the method, and then perform Σ^r on the result.

(III.) When X is of the form $a^x\phi(x)$, where $\phi(x)$ is a rational and integral function of x. Here the formula $f(E)\,a^x\phi(x) = a^x f(aE)\,\phi(x)$ gives us

$$\frac{1}{f(E)}\,a^x\phi(x) = a^x\,\frac{1}{f(aE)}\,\phi(x),$$

which comes under our second rule.

Sin mx and cos mx are really instances of (I.), though the results will be given in an impossible form.

16. Special cases of failure of these rules will occur, as in the analogous cases in differential equations. We shall conclude the Chapter with two examples of this.

Ex. 3. $$(E - a)(E - b)\,u_x = a^x.$$

Here $$f(a) = 0\,;\ \therefore \frac{a^x}{f(a)} = \infty\,.$$

* Its determinateness will serve as our warrant for its truth.

But we may in this case proceed thus:

$$u_x = \frac{a^x}{(E-a)(E-b)} = a^x \frac{1}{(aE-a)\,(aE-b)} \text{ by (III.)}$$

$$= a^x \frac{1}{a\Delta\,(a-b+a\Delta)} = a^{x-1}\frac{x}{a-b+a\Delta},$$

which comes under (II.).

Ex. 4. $$(E-2)^3\,(E-1)\,u^x = x^2 2^x.$$

This will be done in a precisely similar way:

$$u_x = 2^x \frac{x^2}{2^3\Delta^3\,(2E-1)} = 2^{x-3}\Sigma^3 \frac{x^2}{1+2\Delta}$$

$$= 2^{x-3}\Sigma^3\,(x^2-4x+6) = 2^{x-3}\left\{\frac{x^{(5)}}{60} - \frac{x^{(4)}}{8} + x^{(3)}\right\}.$$

17. In a short note in *Tortolini's Annali* (Series I. vol. V.) Maonardi gives a solution of the linear difference-equation with constant coefficients that does not require the preliminary solution of the algebraical equation for E, but the results do not seem of much value.

EXERCISES.

Solve the equations:

1. $u_{x+2} - 3u_{x+1} - 4u_x = m^x$.

2. $u_{x+2} + 4u_{x+1} + 4 = x$.

3. $u_{x+2} + 2u_{x+1} + u_x = x\,(x-1)\,(x-2) + x\,(-1)^x$.

4. $u_{x+2} - 2mu_{x+1} + (m^2+n^2)\,u_x = m^x$.

5. $\Delta u_x + \Delta^2 u_x = x + \sin x$.

6. $u_{x+4} - 6u_{x+2} + 8u_{x+1} - 3u_x = x^2 + (-3)^x$.

7. $\Delta^3 u_x - 5\Delta u_x + 4u_x = 2^x\,(1+\cos x)$.

8. $\Delta^6 u_{x+1} - 2\Delta^6 u_x = x + 3^x$.

9. $u_{x+2} \pm n^2 u_x = \cos mx.$

10. $u_{x+4} \pm 2n^2 u_{x+2} + n^4 u_x = 0.$

11. A person finds his professional income, which for the first year was £a, increase in A.P., the common difference being £b. He saves every year $\frac{1}{m}$ of his income from all sources, laying it out at the end of each year at r per cent. per annum. What will be his income when he has been x years in practice?

12. A seed is planted—when one year old it produces ten-fold, and when two years old and upwards eighteen-fold. Every seed is planted as soon as produced. Find the number of grains at the end of the xth year.

CHAPTER XII.

MISCELLANEOUS PROPOSITIONS AND EQUATIONS. SIMULTANEOUS EQUATIONS.

1. SINCE no class of equations of an order higher than the first have been solved with the completeness which marks the solution of linear difference-equations with constant coefficients, it becomes very important to find what forms of equations can be reduced to this class. The most general case of this reduction is with regard to equations of the form

$$u_{x+n} + A_1\phi(x)\, u_{x+n-1} + A_2\phi(x)\,\phi(x-1)\, u_{x+n-2} + A_3\phi(x)\,\phi(x-1)\,\phi(x-2)\, u_{x+n-3} + \ldots = X \qquad (1),$$

where $A_1\, A_2 \ldots A_n$ are constant, and $\phi(x)$ a known function. These may be reduced to equations with constant coefficients by assuming

$$u_x = \phi(x-n)\,\phi(x-n-1)\ldots\phi(1)\, v_x \qquad (2).$$

For this substitution gives

$$u_{x+n} = \phi(x)\,\phi(x-1)\,\phi(x-2)\ldots\phi(1)\, v_{x+n},$$

$$u_{x+n-1} = \phi(x-1)\,\phi(x-2)\ldots\phi(1)\, v_{x+n-1},$$

and so on; whence substituting and dividing by the common factor $\phi(x)\,\phi(x-1)\ldots\phi(1)$, we get,

$$v_{x+n} + A_1 v_{x+n-1} + A_2 v_{x+n-2} + \ldots = \frac{X}{\phi(x)\,\phi(x-1)\ldots\phi(1)} \qquad (3),$$

an equation with constant coefficients.

In effecting the above transformation we have supposed x to admit of a system of positive integral values. The general transformation would obviously be

$$u_x = \phi(x-n)\,\phi(x-n-1)\ldots\phi(r),$$

r being any particular value of x assumed as *initial.*

Equations of the form

$$u_{x+n} + A_1 a^x u_{x+n-1} + A_2 a^{2x} u_{x+n-2} + \ldots = X,$$

are virtually included in the above class. For, assuming $\phi(x) = a^x$, they may be presented in the form

$$u_{x+n} + A_1\phi(x)\,u_{x+n-1} + A_2 a\,\phi(x)\,\phi(x-1)\,u_{x+n-2} + \ldots = X. \quad (4).$$

Hence, to integrate them it is only necessary to assume

$$u_x = a^{\{1+2+3\ldots+(x-n)\}}\,v_x$$

$$= a^{\frac{(x-n)(x-n+1)}{2}}\,v_x \quad (5).$$

2. By means of the proposition in the last article we can solve all linear binomial equations. Let the equation be

$$u_{x+n} + A_x u_x = B_x. \quad (6).$$

Assume

$$A_x = v_x v_{x-1} \ldots v_{x-n+1} \quad (7).$$

Take logarithms of both sides and let $\log v_{x-n+1} = w_x$, then we have

$$w_{x+n-1} + w_{x+n-2} + \ldots + w_x = \log A_x \quad (8),$$

a linear difference-equation with constant coefficients. Solving this we obtain w_x and thence v_x, which enables us to put (6) into the form

$$U_{x+n} + U_x = X \quad (9)$$

by Art. 1, and thus the equation is solved.

Such equations are however substantially equations of the first degree, and should be treated as such. They state a connection between consecutive members of the series u_r, u_{r+n}, u_{r+2n} ..., and leave these last wholly unconnected with intermediate values of u. We should therefore assume $x = ny$ and the equation would become a linear difference-equation of the first order, the independent variable now proceeding by unit increments.

3. Equations of the form

$$u_{x+1} u_x + a_x u_{x+1} + b_x u_x = c_x \tag{10}$$

can be reduced to linear equations of the second order, and, under certain conditions, to linear equations with constant coefficients*.

Assume

$$u_x = \frac{v_{x+1}}{v_x} - a_x.$$

Then for the first two terms of the proposed equation, we have

$$u_{x+1}(u_x + a_x) = \left(\frac{v_{x+2}}{v_{x+1}} - a_{x+1}\right)\frac{v_{x+1}}{v_x}$$

$$= \frac{v_{x+2}}{v_x} - a_{x+1}\frac{v_{x+1}}{v_x}.$$

Whence substituting and reducing, we find

$$v_{x+2} + (b_x - a_{x+1})\, v_{x+1} - (a_x b_x + c_x)\, v_x = 0 \qquad (11),$$

a linear equation whose coefficients will be constant if the functions $b_x - a_{x+1}$ and $a_x b_x + c_x$ are constant, and which again by the previous section may be reduced to an equation with constant coefficients if those functions are of the respective forms

$$A\phi(x), \quad B\phi(x)\,\phi(x-1).$$

4. Although linear difference-equations with variable coefficients cannot generally be solved, yet, in virtue of their

* Should c_x be zero the equation is at once reduced to a linear equation of the first order by dividing by $u_x u_{x+1}$, and taking $\frac{1}{u_x}$ as our new dependent variable.

linearity, they possess many remarkable properties akin to those possessed by linear differential equations, and which under certain circumstances greatly facilitate their solution. One of these properties is stated in the following Theorem.

THEOREM. *We can depress by unity the order of a linear difference-equation*

$$u_{x+n} + A_x u_{x+n-1} + B_x u_{x+n-2} + \ldots = X \qquad (12),$$

if we know a particular value of u_x *which would satisfy it were the second member* 0.

Let v_x be such a value, so that

$$v_{x+n} + A_x v_{x+n-1} + B_x v_{x+n-2} + \ldots = 0 \qquad (13),$$

and let $u_x = v_x t_x$; then (1) becomes

$$v_{x+n} t_{x+n} + A_x v_{x+n-1} t_{x+n-1} + B_x v_{x+n-2} t_{x+n-2} + \ldots = X.$$

Or $\quad v_{x+n} E^n t_x + A_x v_{x+n-1} E^{n-1} t_x + B_x v_{x+n-2} E^{n-2} t_x \ldots = X.$

Replacing E by $1 + \Delta$, and developing E^n, E^{n-1}, ... in ascending powers of Δ, arrange the result according to ascending differences of t_x. There will ensue

$$(v_{x+n} + A_x v_{x+n-1} + B_x v_{x+n-2} \ldots)\, t_x$$
$$+ P\Delta t_x + Q\Delta^2 t_x \ldots + Z\Delta^n t_x = X.$$

P, Q,...Z being, like the coefficient of t^x, functions of v_x, v_{x+1}, &c. and of the original coefficients A_x, B_x, ...

Now the coefficient of t_x vanishes by (13), whence, making $\Delta t_x = w_x$, we have

$$Pw_x + Q\Delta w_x \ldots + Z\Delta^{n-1} w_x = X,$$

a difference-equation of the $n-1^{\text{th}}$ order for determining w_x*. This being found we have

$$t_x = \Sigma w_x; \quad \therefore u_x = v_x \Sigma w_x.$$

* That the supposition $u_x = v_x t_x$ would lead to a difference-equation of the $(n-1)^{\text{th}}$ order for Δt_x is obvious from *à priori* considerations. For the complementary function of (12) contains a term Cv_x, hence the full value of t_x contains a term C, and thus the full value of Δt_x contains only $n-1$ arbitrary constants, and it must therefore be given by an equation of the $(n-1)^{\text{th}}$ order. That this equation will be linear, follows from the fact that the full value of t_x is *linear in the constants of integration.*

5. We shall demonstrate the Theorem of the last Article by another method, which shews more clearly how the property in question depends on the linearity of the equation; and this second method will teach us how to extend the Theorem to the case in which more than one solution is known.

It was shewn in the last Chapter that linear difference-equations of the nth order had solutions of this form:

$$u_x = C_1 U_x + C_2 V_x + \ldots \quad + I \qquad (14),$$

where C_1, C_2, ... are arbitrary constants, X_1, X_2 are functions of x, and I is a particular integral; also, the part involving the arbitrary constants is the solution of the equation formed by putting 0 for X in (12).

Change x into $x+1$ and eliminate C_1 between the equations, obtaining

$$\left\{E - \frac{U_{x+1}}{U_x}\right\} u_x = C_2 V_x' + \ldots + I' \qquad (15),$$

suppose.

Call $\dfrac{U_{x+1}}{U_x} = M_1$ where M_1 is of course a function of x.

Proceeding as before we shall at length obtain

$$(E - M_n)(E - M_{n-1}) \ldots (E - M_1)\, u_x$$

$$= \text{a quantity depending on } I \text{ alone, and therefore}$$

$$= X \qquad (16),$$

for the left-hand side must be identical with the first member of (12), since, when equated to zero, they have exactly the same solution.

Thus every operation denoted by an operating factor of the form

$$E^n + A_x E^{n-1} + \ldots + N_x$$

can be split up into n consecutive operations, denoted by factors of the form $E - M_r$; and this can be done in many ways, for if we change the order of elimination we shall find that we get wholly different operational factors.

Now suppose we know the first r of the quantities U_x, V_x, then we know the last r operational factors. Assume

$$(E - M_r)(E - M_{r-1})\dots(E - M_1)\,u_x = v_x \qquad (17);$$

then v_x is determined by the equation of the $(n-r)^{\text{th}}$ order,

$$(E - M_n)(E - M_{n-1})\dots(E - M_{r+1}) = X \qquad (18).$$

This last equation we shall now shew can be obtained from our knowledge of U_x, V_x, ...

Let (17) when expanded be

$$(E^r + P_1 E^{r-1} + \dots + P_r)\,u_x = v_x \qquad (19);$$

or, what is the same thing, let the equation whose solution is

$$C_1 U_x + C_2 V_x + \dots C_r Z_x \qquad (20)$$

be
$$(E^r + P_1 E^{r-1} + \dots + P_r)\,u_x = 0.$$

And let (18) when expanded be

$$(E^{n-r} + Q_1 E^{n-r-1} + \dots + Q_{n-r})\,v_x = X \qquad (21),$$

Q_1, Q_2, ... being the coefficients that we are seeking.

Substitute for v_x from (19), we must obtain (13) thereby, and by equating the coefficients of u_x, u_{x+1}... of the resulting equation with their coefficients in (13) we shall obtain n equations for the $n-r$ unknown quantities, Q_1, Q_2.... We shall thus obtain by algebraical solution of these equations the coefficients Q_1, Q_2,...,Q_{n-r}. Thus v_x is made to depend on a linear difference-equation of the $(n-r)^{\text{th}}$ order. When v_x is known, u_x can always be found, for the equation connecting it with v_x is *in its resolved form*, and can thus be solved by successive steps, each consisting of the solution of a linear equation of the first order. If $n-1$ independent solutions be known the equation is reduced to one of the first order, and can therefore be fully solved. Thus we obtain the more general Theorem.

THEOREM*. *We can depress by r the order of a linear difference-equation*

$$u_{x+n} + A_x u_{x+n-1} + B_x u_{x+n-2} + \ldots = X \qquad (22),$$

if we know r independent solutions which would satisfy it were the second member 0; and if we know $n-1$ independent solutions we can solve the equation fully.

Ex. If a solution of

$$u_{x+2} + A_x u_{x+1} + B_x u_x = 0 \qquad (23)$$

be U_x, it is required to solve fully the equation

$$u_{x+2} + A_x u_{x+1} + B_x u_x = X \qquad (24).$$

By the last Article equation (23) must be of the form

$$(E - P_x)\left(E - \frac{U_{x+1}}{U_x}\right) u_x = 0 \qquad (25);$$

and on comparing the two forms we obtain $P_x \cdot \frac{U_{x+1}}{U_x} = B_x$, and therefore (24) may be written

$$\left(E - B_x \frac{U_x}{U_{x+1}}\right)\left(E - \frac{U_{x+1}}{U_x}\right) u_x = X \qquad (26).$$

The first step in the solution gives us

$$\left(E - \frac{U_{x+1}}{U_x}\right) u_x = \Pi\left(B_{x-1} \frac{U_{x-1}}{U_x}\right)\left[\Sigma \frac{X}{\Pi\left(B_x \frac{U_x}{U_{x+1}}\right)} + C\right]$$

$$= \frac{U_{r-1}}{U_x} \Pi(B_{x-1})\left[\Sigma \frac{U_{x+1} X}{U_r \Pi(B_x)} + C\right] \qquad (27).$$

Dividing by U_{x+1}, summing, and multiplying by U_x, we obtain

$$u_x = U_x \Sigma\left\{\frac{U_{r-1}}{U_x U_{x+1}} \Pi(B_{x-1})\left[\Sigma \frac{U_{x+1} X}{U_r \Pi(B_x)} + C\right]\right\} + C' U_x \qquad (28).$$

* Tardi gives a proof of this theorem (*Tortolini*, Series I. vol. I.), and especially considers the latter case.

6. Certain forms of linear equations can be solved by performing Δ upon them one or more times.

Take, for example, the equation

$$(a+bx)\,\Delta^2 u_x + (c+dx)\,\Delta u_x + e u_x = 0 \qquad (29)$$

and perform Δ^n upon it. By the formula at the top of page 21,

$$\Delta^n u_x v_x = (E\Delta' + \Delta)^n u_x v_x,$$

we have

$$\{a+b\,(x+n)\}\,\Delta^{n+2}u_x + nb\Delta^{n+1}u_x + \{c+d\,(x+n)\}\,\Delta^{n+1}u_x + nd\Delta^n u_x + e\Delta^n u_x = 0 \quad (30);$$

and if we take $n = -\frac{e}{d}$, supposing that to be an integer, we have a linear equation of the first order for $\Delta^{n+1}u_x$.

Ex. $$x\Delta^2 u_x + (x-2)\,\Delta u_x - u_x = 0 \qquad (31).$$

Performing Δ on it we have

$$(x+1)\,\Delta^3 u_x + x\Delta^2 u_x = 0,$$

which gives

$$\Delta^2 u_x = \frac{c}{|\underline{x}} \qquad (32);$$

$$\therefore \Delta u_x = \Sigma \frac{c}{|\underline{x}} + c' \qquad (33).$$

Substituting from (32) and (33) in (31) we obtain

$$u_x = \frac{c}{|\underline{x-1}} + (x-2)\left\{\Sigma \frac{c}{|\underline{x}} + c'\right\} \qquad (34).$$

A more general form of this solution would be

$$u_x = \frac{c}{\Gamma(x)} + (x-2)\left\{\Sigma \frac{c}{\Gamma(x+1)} + c'\right\} \qquad (35).$$

The method is due to Bronwin (*Camb. Math. Jour.* Vol. III. and *Camb. and Dub. Math. Jour.* Vol. II.).

7. The solution of two very remarkable non-linear equations has been deduced by Prof. Sylvester from that of linear equations with constant coefficients.

Let
$$u_{x+n} + p_1 u_{x+n-1} + \dots + p_n u_x = 0 \qquad (36)$$
be any such equation. Then writing it down for the next n values of x

$$u_{x+n+1} + p_1 u_{x+n} + \dots + p_n u_{x+1} = 0,$$
$$\dots \qquad = 0,$$
$$u_{x+2n} + p_1 u_{x+2n-1} + \dots + p_n u_{x+n} = 0.$$

Eliminating the quantities $p_1, p_2, \dots$ we obtain

$$\begin{vmatrix} u_{x+n}, & u_{x+n-1}, \dots & u_x \\ u_{x+n+1} & \dots\dots & u_{x+1} \\ \dots & \dots & \dots \\ u_{x+2n} & \dots\dots & u_{x+n} \end{vmatrix} = 0 \qquad (37),$$

an equation which must be satisfied by every solution of (36).

Now the solution of (36) is
$$u_x = A\alpha^x + B\beta^x + \dots \text{ to } n \text{ terms} \qquad (38),$$
where A, B, ... are arbitrary and α, β, ... depend on $p_1, p_2, \dots$ and these last do not appear in equation (37) which we are now considering. Hence (38) will be the solution of (37), α, β, ... being *also* considered arbitrary, thus making the full number of $2n$ arbitrary constants.

By a slight variation in the method of elimination we can obtain the solution of a yet more general equation. Taking the last term of each of the equations to the other side and eliminating $p_1, p_2, \dots p_{n-1}$, we obtain

$$\begin{vmatrix} u_{x+n} \dots\dots & u_{x+1} \\ u_{x+n+1} \dots\dots & \\ \dots\dots & \\ u_{x+2n-1} \dots & u_{x+n} \end{vmatrix} = -p_n \begin{vmatrix} u_x, & u_{x+n-1} \dots\dots & u_{x+1} \\ u_{x+1}, & \dots\dots & \\ \dots & \dots\dots & \\ u_{x+n-1}, & \dots\dots & u_{x+n} \end{vmatrix}$$

$$= (-1)^n p_n \begin{vmatrix} u_{x+n-1}, & u_{x+n-2}, \dots\dots & u_x \\ u_{x+n} & \dots\dots & u_{x+1} \\ \dots & \dots\dots & \dots \\ u_{x+2n-2} & \dots\dots & u_{x+n-1} \end{vmatrix} \qquad (39),$$

or calling the last determinant P_x

$$P_{x+1} = (-1)^n p_n P_x. \qquad (40),$$

the solution of which may be written

$$P_x = C\{(-1)^n p_n\}^x \qquad (41).$$

Thus the solution of the equation (writing $n+1$ for n)

$$\begin{vmatrix} u_{x+n} \ldots\ldots\ldots\ldots u_x \\ \ldots\ldots\ldots\ldots\ldots\ldots \\ u_{x+2n-1} \ldots\ldots\ldots\ldots \\ u_{x+2n}, u_{x+2n-1}, \ldots u_{x+n} \end{vmatrix} = Cm^x. \qquad (42)$$

is $$u_x = A\alpha^x + B\beta^x + \ldots \text{ to } n+1 \text{ terms} \qquad (43),$$

where A, B, ... and α, β, ... are arbitrary constants limited by the two equations of condition

$$m = \alpha\beta\gamma\ldots$$

and $\qquad C =$ the determinant P for some value of x.

If we take this last-named value to be zero, it is evident that

$$C = \begin{vmatrix} A\alpha^n \ldots\ldots\ldots\ldots \\ A\alpha^2 \ldots\ldots\ldots\ldots \\ A\alpha, B\beta, C\gamma, \ldots \\ A, B, C, \ldots\ldots \end{vmatrix} \times \begin{vmatrix} 1, 1, 1, \ldots\ldots \\ \alpha, \beta, \gamma, \ldots\ldots \\ \ldots\ldots\ldots\ldots \\ \end{vmatrix}$$

$$= ABC\ldots \begin{vmatrix} 1, 1, 1, \ldots\ldots \\ \alpha, \beta, \gamma, \ldots\ldots \\ \ldots\ldots\ldots\ldots \\ \alpha^n \ldots\ldots\ldots \end{vmatrix}^2 \times (-1)^{\frac{n(n-1)}{2}}$$

$= ABC \ldots$ product of squares of differences of $\alpha, \beta, \gamma\ldots$ taken with the proper sign.

Ex. The equation

$$u_x u_{x+2} - u_x^{\,2} = C \qquad (44)$$

may be supposed to be derived from the equation

$$u_{x+1} + pu_x = -u_{x-1},$$

which gives also

$$u_{x+2} + pu_{x+1} = -u_x.$$

Whence eliminating p we have

$$u_{x+1}^2 - u_x u_{x+2} = u_x^2 - u_{x-1} u_{x+1} \text{ and } \therefore = \text{constant},$$

since it is equal to its consecutive value.

Hence $$u_x = A\alpha^x + B\beta^x, \text{ where } \alpha\beta = 1,$$

and $$(A\alpha^{-1} + B\beta^{-1})(A\alpha + B\beta) - (A+B)^2 = C;$$

$$\therefore AB\alpha^2 + AB\beta^2 - 2AB\alpha\beta = C\alpha\beta,$$

$$\text{or } C = AB(\alpha - \beta)^2.$$

Simultaneous Equations.

8. Instead of a single equation involving one function we may find that we have a system of n equations involving n unknown functions of the independent variable. The method by which we reduce this to the former case is so obvious that we shall not dwell upon it. We must by the performance of Δ or E obtain a system of derived equations sufficient to enable us by elimination to deduce a final equation involving only one of the variables with its differences and successive values. The integrations of this will give the general value of that variable, and the equations employed in the process of elimination will enable us to express each other dependent variable by means of it. If the coefficients are constant we may simply separate the symbols and effect the eliminations as if those symbols were algebraic.

Ex. 1. $$\left.\begin{aligned} u_{x+1} - a^2 x v_x &= 0 \\ v_{x+1} - \quad x u_x &= 0 \end{aligned}\right\}.$$

From the first we have

$$u_{x+2} - a^2(x+1)\, v_{x+1} = 0.$$

Hence eliminating v_{x+1} by the second

$$u_{x+2} - a^2 x(x+1)\, u_x = 0,$$

the solution of which is

$$u = \underline{|x-1}\, \{Ca^x + C'(-a)^x\},$$

and by the first equation

$$v_x = \frac{u_{x+1}}{a^2 x} = \frac{|x-1}{a^2}\{Ca^{x+1} + C'(-a)^{x+1}\}.$$

Ex. 2.
$$\left.\begin{aligned} u_{x+1} + 2v_{x+1} - u_x &= 0 \\ v_{x+1} - 2u_x - v_x &= a^x \end{aligned}\right\}.$$

This may be written

$$(E-1)\,u_x + 2Ev_x = 0,$$
$$-2u_x + (E-1)\,v_x = a^x;$$
$$\therefore \{(E-1)^2 + 4E\}\,u_x = -2Ea^x;$$
$$\text{or } (E+1)^2\,u_x = -2a^{x+1}.$$

This gives

$$u_x = (C + C'x)(-1)^x - \frac{2a^{x+1}}{(a+1)^2},$$

and from the first equation

$$2v_{x+1} = -\Delta u_x = \{2C + C'(2x+1)\}(-1)^x + \frac{2a^{x+1}(a-1)}{(a+1)^2};$$

$$\therefore v_x = -\left\{C - \frac{C'}{2} + C'x\right\}(-1)^x + \frac{a^x(a-1)}{(a+1)^2}.$$

9. On the subject of linear equations with variable coefficients the student should see a remarkable paper by Christoffel (*Crelle*, LV. 281), in which he dwells on the anomalies produced by the passage through a value which causes the coefficient of the first or last term to vanish. On the condition that an expression in differences should be capable of immediate summation, i.e. should be analogous to an exact differential, see Minich, (*Tortolini*, Series I. vol. I. 321).

EXERCISES.

Integrate the equations

1. $u_{x+2} - xu_{x+1} + (x-1)\,u_x = \sin x$, one portion of the complementary function being a constant.

2. $u_{x+n-1} + u_{x+n-2} + \ldots + u_x = 0.$

3. $u_x = x\,(u_{x-1} + u_{x-2}).$

4. $u_{x+1} = x\,(u_x + u_{x-1}).$

5. $u_{x+2} - 2(x-1)\,u_{x+1} + (x-1)(x-2)\,u_x = x.$

6. $u_{x+2}\,u_{x+1}^{2n}\,u_x^{n^2} = a.$

7. $(x+3)^2\,u_{x+2} - \dfrac{2(x+2)^2}{x+1}\,u_{x+1} + \dfrac{(x+1)^2(x+2)}{x}\,u_x = 0.$

8. Integrate the simultaneous equations

$$\left.\begin{aligned} u_{x+1} - v_x &= 2m(x+1) \\ v_{x+1} - u_x &= -2m(x+1) \end{aligned}\right\}.$$

9.

$$\left.\begin{aligned} u_{x+1} + (-1)^x v_x &= 0 \\ v_{x+1} + (-1)^x u_x &= 0 \end{aligned}\right\}.$$

10.

$$\left.\begin{aligned} v_{x+1} - u_x &= (l-m)\,x \\ w_{x+1} - v_x &= (m-n)\,x \\ u_{x+1} - w_x &= (n-l)\,x \end{aligned}\right\}.$$

11.

$$\left.\begin{aligned} u_{x+2} + 2v_{x+1} - 8u_x &= a^x \\ v_{x+2} - u_{x+1} - 2v_x &= a^{-x} \end{aligned}\right\}.$$

12. When the solution of a non-linear equation of the first order is made to depend upon that of a linear equation of the second order whose second member is 0 by assuming

$$u_x = \frac{v_{x+1}}{v_x} - a_x$$

(Art. 3), shew that the two constants which appear in the value of v_x effectively produce only one in that of u_x.

13. The equation

$$u_{x+2} - (a^{x+1} + a^{-x})\,u_{x+1} + u_x = 0$$

may be resolved into two equations of differences of the first order.

14. Given that a particular solution of the equation

$$u_{x+2} - a(a^x+1)\,u_{x+1} + a^{x+1}\,u_x = 0 \text{ is } u_x = ca^{\frac{x(x-1)}{2}},$$

deduce the general solution, and also shew that the above equation may be solved without the previous knowledge of a particular integral.

15. The equation

$$u_x u_{x+1} u_{x+2} = a\,(u_x + u_{x+1} + u_{x+2})$$

may be integrated by assuming $u_x = \sqrt{a}\tan v_x$.

16. Shew also that the general integral of the above equation is included in that of the equation $u_{x+3} - u_x = 0$, and hence deduce the former.

17. Shew how to integrate the equation

$$u_{x+1}u_{x+2} + u_{x+2}u_x + u_x u_{x+1} = m^2.$$

18. Solve the equations

$$\left.\begin{aligned} u_{x+1} &= (n - m^2)\, v_x + u_x, \\ v_{x+1} &= (2m+1)\, v_x + u_x, \end{aligned}\right\}$$

and shew that if m be the integral part of $\sqrt{n}$, $\frac{u_x}{v_x}$ converges as x increases to the decimal part of $\sqrt{n}$.

19. If a_1 be a fourth proportional to a, b, c, b_1 a fourth proportional to b, c, a, and c_1 to c, a, b, and a_2, b_2, c_2 depend in the same manner on a_1, b_1, c_1, find the linear equation of differences on which a_n depends and solve it.

20. Solve the equation

$$x\,(x+1)\,\Delta^2 u_x + k\,(1-x)\,\Delta u_x + k u_x = 0.$$

21. Solve the equation $\begin{vmatrix} u_{x+5}, & u_{x+4}, & u_{x+3} \\ u_{x+4}, & u_{x+3}, & u_{x+2} \\ u_{x+3}, & u_{x+2}, & u_{x+1} \end{vmatrix} = C,$

considering specially the case when C is zero.

22. If $v_0, v_1, v_2, \ldots$ be a series of quantities the successive terms of which are connected by the general relation

$$v_{m+1} = v_1 v_m - v_{m-1},$$

and if v_0, v_1 be any given quantities, find the value of v_n. [S.P.]

23. If n integers are taken at random and multiplied together in the denary scale, find the chance that the figure in the unit's place will be 2.

24. Shew that a solution of the equation

$$u_{x+n}u_{x+n-1}\dots u_x = a\left(u_{x+n} + u_{x+n-1} + \dots u\right)$$

is included in that of

$$u_{x+n+1} - u_x = 0,$$

and is consequently

$$u_x = C_1\alpha^x + C_2\alpha^{2x} + \dots + C_{n+1}\alpha^{(n+1)x},$$

where α is one of the imaginary $(n+1)^{\text{th}}$ roots of unity, the $n+1$ constants being subject to an equation of condition.

25. Solve the equation

$$P_{n+1} = P_n + P_{n-1}P_3 + P_{n-2}P_4 + \dots + P_3P_{n-1} + P_n,$$

and shew that it is equivalent to

$$P_{n+1} = \frac{4n-6}{n}P_n.$$

[Catalan, *Liouville*, III. 508.]

26. Shew that

$$u_{x+1} = \frac{u_x}{x} + u_{x-1}$$

can be satisfied by $u_{2x} = u_{2x+1}$ or u_{2x-1}, and that thus its solution is

$$u_{2x} = C.\frac{3.5.7\dots(2x-1)}{2.4.6\dots(2x-2)} + C'.\frac{2.4.6\dots 2x}{1.3.5\dots(2x-1)},$$

$$u_{2x-1} = C.\frac{3.5.7\dots(2x-1)}{2.4.6\dots(2x-2)} + C'.\frac{2.4.6\dots(2x-2)}{1.3.5\dots(2x-3)},$$

and deduce therefrom the solution of

$$u_{x+1} = u_x + (x^2 - x)\,u_{x-1}.$$

[Sylvester, *Phil. Mag.*]

CHAPTER XIII.

LINEAR EQUATIONS WITH VARIABLE COEFFICIENTS. SYMBOLICAL AND GENERAL METHODS.

1. THE symbolical methods for the solution of differential equations whether in finite terms or in series (*Diff. Equations*, Chap. XVII.) are equally applicable to the solution of difference-equations. Both classes of equations admit of the same symbolical form, the elementary symbols combining according to the same ultimate laws. And thus the only remaining difference is one of interpretation, and of processes founded upon interpretation. It is that kind of difference which exists between the symbols $\left(\frac{d}{dx}\right)^{-1}$ and Σ.

It has been shewn that if in a linear differential equation we assume $x = \epsilon^{\theta}$, the equation may be reduced to the form

$$f_0\left(\frac{d}{d\theta}\right) u + f_1\left(\frac{d}{d\theta}\right) \epsilon^{\theta} u + f_2\left(\frac{d}{d\theta}\right) \epsilon^{2\theta} u \ldots + f_n\left(\frac{d}{d\theta}\right) \epsilon^{n\theta} u = U, \tag{1}$$

U being a function of θ. Moreover, the symbols $\frac{d}{d\theta}$ and ϵ^{θ} obey the laws,

$$\left.\begin{aligned} f\left(\frac{d}{d\theta}\right) \epsilon^{m\theta} u &= \epsilon^{m\theta} f\left(\frac{d}{d\theta} + m\right) u \\ f\left(\frac{d}{d\theta}\right) \epsilon^{m\theta} &= f(m)\, \epsilon^{m\theta} \end{aligned}\right\} \tag{2}$$

And hence it has been shewn to be possible, 1st, to express the solution of (1) in series, 2ndly, to effect by general theorems the most important transformations upon which finite integration depends.

Now $\frac{d}{d\theta}$ and ϵ^θ are the equivalents of $x\frac{d}{dx}$ and x, and it is proposed *to develope in this chapter the corresponding theory of difference-equations founded upon the analogous employment of the symbols* $x\frac{\Delta}{\Delta x}$ *and* xE, *supposing* Δx *arbitrary, and therefore*

$$\Delta\phi(x) = \phi(x + \Delta x) - \phi(x),$$
$$E\phi(x) = \phi(x + \Delta x).$$

PROP. 1. *If the symbols* π *and* ρ *be defined by the equations*

$$\pi = x\frac{\Delta}{\Delta x}, \quad \rho = xE \qquad (3),$$

they will obey the laws

$$\left.\begin{aligned} f(\pi)\,\rho^m u &= \rho^m f(\pi + m)\,u \\ f(\pi)\,\rho^m &= f(m)\,\rho^m \end{aligned}\right\} \qquad (4),$$

the subject of operation in the second theorem being unity.

1st. Let $\Delta x = r$, and first let us consider the interpretation of $\rho^m u_x$.

Now $$\rho u_x = xEu_x = xu_{x+r};$$
$$\therefore \rho^2 u_x = \rho x u_{x+r} = x(x+r)\,u_{x+2r},$$
whence generally

$$\rho^m u_x = x(x+r)\ldots\{x+(m-1)\,r\}\,u_{x+mr},$$

an equation to which we may also give the form

$$\rho^m u_x = x(x+r)\ldots\{x+(m-1)\,r\}\,E^m u_x \qquad (5).$$

If $u_x = 1$, then, since $u_{x+mr} = 1$, we have

$$\rho^m 1 = x(x+r)\ldots\{x+(m-1)\,r\},$$

to which we shall give the form

$$\rho^m = x(x+r)\ldots\{x+(m-1)\,r\},$$

the subject 1 being understood.

2ndly. Consider now the series of expressions

$$\pi\rho^m u_x, \quad \pi^2\rho^m u_x, \ldots \pi^n\rho^m u_x.$$

Now

$$\pi\rho^m u_x = x\frac{\Delta}{\Delta x}x(x+r)\ldots\{x+(m-1)r\}u_{x+mr}$$

$$= x\frac{(x+r)\ldots(x+mr)u_{x+(m+1)r} - x\ldots\{x+(m-1)r\}u_{x+mr}}{r}$$

$$= x\ldots\{x+(m-1)r\}\frac{(x+mr)u_{x+(m+1)r} - xu_{x+mr}}{r}$$

$$= x\ldots\{x+(m-1)r\}E^m\frac{xu_{x+r} - (x-mr)u_x}{r}$$

$$= \rho^m\frac{xu_{x+r} - (x-mr)u_x}{r}, \text{ by (5)},$$

$$= \rho^m\left(x\frac{u_{x+r} - u_x}{r} + mu_x\right)$$

$$= \rho^m\left(x\frac{\Delta}{\Delta x}u_x + mu_x\right)$$

$$= \rho^m(\pi+m)u_x.$$

Hence

$$\pi^2\rho^m u_x = \pi\rho^m(\pi+m)u_x$$
$$= \rho^m(\pi+m)^2u_x,$$

and generally

$$\pi^n\rho^m u_x = \rho^m(\pi+m)^n u_x.$$

Therefore supposing $f(\pi)$ a function expressible in ascending powers of π, we have

$$f(\pi)\rho^m u = \rho^m f(\pi+m)u \qquad (6),$$

which is the first of the theorems in question.

Again, supposing $u = 1$, we have

$$f(\pi)\rho^m 1 = \rho^m f(\pi+m)1$$
$$= \rho^m\left\{f(m) + f'(m)\pi + \frac{f''(m)}{1.2}\pi^2 + \ldots\right\}1.$$

But $\pi 1 = x \frac{\Delta}{\Delta x} 1 = 0$, $\pi^2 1 = 0$, ... Therefore

$$f(\pi)\rho^m 1 = \rho^m f(m) 1.$$

Or, omitting but leaving understood the subject unity,

$$f(\pi)\rho^m = f(m)\rho^m \qquad (7).$$

PROP. 2. *Adopting the previous definitions of π and ρ, every linear difference-equation admits of symbolical expression in the form*

$$f_0(\pi)u_x + f_1(\pi)\rho u_x + f_2(\pi)\rho^2 u_x \ldots + f_m(\pi)\rho^m u_x = X \qquad (8).$$

The above proposition is true irrespectively of the particular value of Δx, but the only cases which it is of any importance to consider are those in which $\Delta x = 1$ and -1.

First suppose the given difference-equation to be

$$X_0 u_{x+n} + X_1 u_{x+n-1} \ldots + X_n u_x = \phi(x) \qquad (9).$$

Here it is most convenient to assume $\Delta x = 1$ in the expressions of π and ρ. Now multiplying each side of (9) by

$$x(x+1)\ldots(x+n-1),$$

and observing that by (5)

$$xu_{x+1} = \rho u_x, \quad x(x+1)u_{x+2} = \rho^2 u_x, \ \ldots,$$

we shall have a result of the form

$$\phi_0(x)u_x + \phi_1(x)\rho u_x \ldots + \phi_n(x)\rho^n u_x = \phi_1(x) \qquad (10).$$

But since $\Delta x = 1$,

$$\pi = x\Delta, \quad \rho = xE$$
$$= x\Delta + x.$$

Hence

$$x = -\pi + \rho,$$

and therefore

$$\phi_0(x) = \phi_0(-\pi + \rho), \quad \phi_1(x) = \phi_1(-\pi + \rho), \ \ldots.$$

These must be expressed in ascending powers of ρ, regard being paid to the law expressed by the first equation of (4).

The general theorem for this purpose, though its application can seldom be needed, is

$$F_0(\pi-\rho) = F_0(\pi) - F_1(\pi)\,\rho + F_2(\pi)\frac{\rho^2}{1.2}$$

$$- F_3(\pi)\frac{\rho^3}{1.2.3} + \dots \qquad (11),$$

where $F_1(\pi)$, $F_2(\pi)$, ..., are formed by the law

$$F_m(\pi) = F_{m-1}(\pi) - F_{m-1}(\pi-1).$$

(*Diff. Equations*, p. 439.)

The equation (10) then assumes after reduction the form (8).

Secondly, suppose the given difference-equation presented in the form

$$X_0 u_x + X_1 u_{x-1} \dots + X_n u_{x-n} = X \qquad (12).$$

Here it is most convenient to assume $\Delta x = -1$ in the expression of π and ρ.

Now multiplying (12) by $x(x-1)\dots(x-n+1)$, and observing that by (5)

$$xu_{x-1} = \rho u_x, \quad x(x-1)\,u_{x-2} = \rho^2 u_x, \ \dots,$$

the equation becomes

$$\phi_0(x)\,u_x + \phi_1(x)\,\rho u_x \dots + \phi_n(x)\,\rho^n u_x = X,$$

but in this case as is easily seen we have

$$x = \pi + \rho,$$

whence, developing the coefficients, if necessary, by the theorem

$$F_0(\pi+\rho) = F_0(\pi) + F_1(\pi)\,\rho + F_2(\pi)\frac{\rho^2}{1.2} + \dots \qquad (13),$$

where as before

$$F_m(\pi) = F_{m-1}(\pi) - F_{m-1}(\pi-1),$$

we have again on reduction an equation of the form (8).

2. It is not always necessary in applying the above methods of reduction to multiply the given equation by a factor of the form

$$x(x+1)\ldots(x+n-1), \text{ or } x(x-1)\ldots(x-n+1),$$

to prepare it for the introduction of ρ. It may be that the constitution of the original coefficients $X_0, X_1 \ldots X_n$ is such as to render this multiplication unnecessary; or the requisite factors may be introduced in another way. Thus resuming the general equation

$$X_0 u_x + X_1 u_{x-1} \ldots + X_n u_{x-n} = 0 \qquad (14),$$

assume

$$u_x = \frac{v_x}{1.2\ldots x}.$$

We find

$$X_0 v_x + X_1 x v_{x-1} \ldots + X_n x(x-1)\ldots(x-n+1)\, v_{x-n} = 0 \quad (15).$$

Hence assuming

$$\pi = x\frac{\Delta}{\Delta x}, \quad \rho = xE,$$

where $\Delta x = -1$, we have

$$X_0 v_x + X_1 \rho v_x \ldots + X_n \rho^n v_x = 0 \qquad (16),$$

and it only remains to substitute $\pi + \rho$ for x and develope the coefficients by (13).

3. A preliminary transformation which is often useful consists in assuming $u_x = \mu^x v_x$. This converts the equation

$$X_0 u_x + X_1 u_{x-1} \ldots + X_n u_{x-n} = 0 \qquad (17)$$

into

$$\mu^n X_0 v_x + \mu^{n-1} X_1 v_{x-1} \ldots X_n v_{x-n} = 0 \qquad (18),$$

putting us in possession of a disposable constant μ.

4. When the given difference-equation is expressed directly in the form

$$X_0 \Delta^n u + X_1 \Delta^{n-1} u \ldots + X_n u = 0 \qquad (19),$$

it may be convenient to apply the following theorem.

Theorem. If $\pi = x\dfrac{\Delta}{\Delta x}$, $\rho = xE$, then

$$\pi(\pi-1)\ldots(\pi-n+1)u = x(x+\Delta x)\ldots$$

$$\{x+(n-1)\Delta x\}\left(\frac{\Delta}{\Delta x}\right)^n u \qquad (20).$$

To prove this we observe that since

$$F(\pi)\rho^n u = \rho^n F(\pi+n)u,$$

therefore $\qquad F(\pi+n)u = \rho^{-n}F(\pi)\rho^n u,$

whence $\qquad F(\pi-n)u = \rho^n F(\pi)\rho^{-n}u.$

Now reversing the order of the factors $\pi, \pi-1, \ldots \pi-n+1$ in the first member of (20), and applying the above theorem to each factor separately, we have

$$(\pi-n+1)(\pi-n+2)\ldots\pi u$$
$$=\rho^{n-1}\pi\rho^{-n+1}\rho^{n-2}\pi\rho^{-n+2}\ldots\pi u$$
$$=\rho^n(\rho^{-1}\pi)^n u.$$

But $\rho^{-1}\pi = (xE)^{-1}x\dfrac{\Delta}{\Delta x} = E^{-1}x^{-1}x\dfrac{\Delta}{\Delta x} = E^{-1}\dfrac{\Delta}{\Delta x}$;

$$\therefore\ (\pi-n+1)(\pi-n+2)\ldots\pi = \rho^n\left(E^{-1}\frac{\Delta}{\Delta x}\right)^n$$

$$= \rho^n E^{-n}\left(\frac{\Delta}{\Delta x}\right)^n.$$

But $\rho^n u = x(x+r)\ldots\{x+(n-1)r\}E^n u$, whence

$$(\pi-n+1)(\pi-n+2)\ldots\pi u = x(x+r)\ldots\{x+(n-1)r\}\left(\frac{\Delta}{\Delta x}\right)^n u,$$

which, since $r = \Delta x$, agrees with (20).

When $\Delta x = 1$, the above gives

$$\pi(\pi-1)\ldots(\pi-n+1) = x(x+1)\ldots(x+n-1)\Delta^n \qquad (21).$$

Hence, resuming (19), multiplying both sides by

$$x(x+1)\ldots(x+n-1),$$

and transforming, we have a result of the form

$$\phi_0 (x) \pi (\pi - 1) \dots (\pi - n + 1) u$$
$$+ \phi_1 (x) \pi (\pi - 1) \dots (\pi - n + 2) u + \dots = 0.$$

It only remains then to substitute $x = -\pi + \rho$, develope the coefficients, and effect the proper reductions.

Solution of Linear Difference-Equations in series.

5. Supposing the second member 0, let the given equation be reduced to the form

$$f_0 (\pi) u + f_1 (\pi) \rho u + f_2 (\pi) \rho^2 u \dots + f_n (\pi) \rho^n u = 0 \qquad (22),$$

and assume $u = \Sigma a_m \rho^m$. Then substituting, we have

$$\Sigma \{f_0 (\pi) a_m \rho^m + f_1 (\pi) a_m \rho^{m+1} \dots + f_n (\pi) a_m \rho^{m+n}\} = 0,$$

whence, by the second equation of (4),

$$\Sigma \{f_0 (m) a_m \rho^m + f_1 (m+1) a_m \rho^{m+1} \dots + f_n (m+n) a_m \rho^{m+n}\} = 0,$$

in which the aggregate coefficient of ρ^m equated to 0 gives

$$f_0 (m) a_m + f_1 (m) a_{m-1} \dots + f_n (m) a_{m-n} = 0 \qquad (23).$$

This, then, is the relation connecting the successive values of a_m. The lowest value of m, corresponding to which a_m is arbitrary, will be determined by the equation

$$f_0 (m) = 0,$$

and there will thus be as many values of u expressed in series as the equation has roots.

If in the expression of π and ρ we assume $\Delta x = 1$, then since

$$\rho^m = x (x+1) \dots (x+m-1) \qquad (24),$$

the series $\Sigma a_m \rho^m$ will be expressed in ascending factorials of the above form. But if in expressing π and ρ we assume $\Delta x = -1$, then since

$$\rho^m = x(x-1)\ldots(x-m+1) \qquad (25),$$

the series will be expressed in factorials of the latter form.

Ex. 1. Given

$$(x-a)\,u_x - (2x-a-1)\,u_{x-1} + (1-q^2)(x-1)\,u_{x-2} = 0;$$

required the value of u_x in descending factorials.

Multiplying by x, and assuming $\pi = x\dfrac{\Delta}{\Delta x}$, $\rho = xE$, where $\Delta x = -1$, we have

$$x(x-a)\,u_x - (2x-a-1)\,\rho u_x + (1-q^2)\,\rho^2 u_x = 0,$$

whence, substituting $\pi + \rho$ for x, developing by (13), and reducing,

$$\pi(\pi - a)\,u_x - q^2\rho^2 u_x = 0. \qquad (a).$$

Hence
$$u_x = \Sigma a_m \rho^m,$$

the initial values of a_m corresponding to $m = 0$ and $m = a$ being arbitrary, and the succeeding ones determined by the law

$$m(m-a)\,a_m - q^2 a_{m-2} = 0.$$

Thus we have for the complete solution

$$u_x = C\left\{1 + \frac{q^2 x^{(2)}}{2(2-a)} + \frac{q^4 x^{(4)}}{2\,.\,4\,.\,(2-a)\,.\,(4-a)} + \ldots\right\}$$

$$+ C'\left\{x^{(a)} + \frac{q^2 x^{(a+2)}}{2\,.\,(2+a)} + \frac{q^4 x^{(a+4)}}{2\,.\,4\,(2+a)(4+a)} + \ldots\right\} \qquad (b).$$

It may be observed that the above difference-equation might be so prepared that the complete solution should admit of expression in finite series. For assuming $u_x = \mu^x v_x$, and then transforming as before, we find

$$\mu^2\pi(\pi - a)\, v_x + (\mu^2 - \mu)(2\pi - a - 1)\, \rho v_x$$
$$+ \{(\mu - 1)^2 - q^2\}\, \rho^2 v_x = 0 \qquad (c),$$

which becomes binomial if $\mu = 1 \pm q$, thus giving

$$\pi(\pi - a)\, v_x + \frac{\mu - 1}{\mu}(2\pi - a - 1)\, \rho v_x = 0.$$

Hence we have for either value of μ,

$$u_x = \mu^x \Sigma a_m \rho^m = \mu^x \Sigma a_m x(x-1)\ldots(x-m+1) \qquad (d),$$

the initial value of m being 0 or a, and all succeeding values determined by the law

$$m(m-a)\, a_m + \frac{\mu - 1}{\mu}(2m - a - 1)\, a_{m-1} = 0 \qquad (e).$$

It follows from this that the series in which the initial value of m is 0 terminates when a is a positive odd number, and the series in which the initial value of m is a terminates when a is a negative odd number. Inasmuch however as there are two values of μ, either series, by giving to μ both values in succession, puts us in possession of the complete integral.

Thus in the particular case in which a is a positive odd number we find

$$u_x = C(1+q)^x \left\{1 - \frac{q}{1+q}\, \frac{(1-a)\, x}{1\,.\,(1-a)} + \frac{q^2}{(1+q)^2}\, \frac{(1-a)(3-a)\, x^{(2)}}{1\,.\,2\,(1-a)(2-a)} - \ldots\right\}$$

$$+ C'(1-q)^x \left\{1 + \frac{q}{1+q}\, \frac{(1-a)\, x}{1\,.\,(1-a)} + \frac{q^2}{(1+q)^2}\, \frac{(1-a)(3-a)\, x^{(2)}}{1\,.\,2\,(1-a)(3-a)} + \ldots\right\} \qquad (k).$$

The above results may be compared with those of p. 454 of *Differential Equations*.

Finite solution of Difference-Equations.

6. The simplest case which presents itself is when the symbolical equation (8) is monomial, i.e. of the form

$$f_0(\pi)\, u = X \qquad (26).$$

We have thus

$$u = \{f_0(\pi)\}^{-1} X \qquad (27).$$

Resolving then $\{f_0(\pi)\}^{-1}$ as if it were a rational algebraic fraction, the complete value of u will be presented in a series of terms of the form

$$A\,(\pi - a)^{-i} X.$$

But by (4) we have

$$(\pi - a)^{-i} X = \rho^a (\pi)^{-i} \rho^{-a} X \qquad (28).$$

It will suffice to examine in detail the case in which $\Delta x = 1$ in the expression of π and ρ.

To interpret the second member of (28) we have then

$$\rho^a \phi(x) = x(x+1) \dots (x+a-1)\, \phi(x+a),$$

$$\rho^{-a} \phi(x) = \frac{\phi(x-a)}{(x+1)(x+2)\dots(x+a)},$$

$$\pi^{-i} \phi(x) = (x\Delta)^{-i} \phi(x)$$

$$= \Sigma \frac{1}{x} \Sigma \frac{1}{x} \dots \phi(x);$$

the complex operation $\Sigma \frac{1}{x}$, denoting division of the subject by x and subsequent integration, being repeated i times.

Should X however be rational and integral it suffices to express it in factorials of the forms

$$x, \quad x(x+1), \quad x(x+1)(x+2), \dots$$

to replace these by ρ, ρ^2, ρ^3, ... and then interpret (27) at once by the theorem

$$\{f_0(\pi)\}^{-1}\rho^m = \{f_0(m)\}^{-1}\rho^m$$
$$= \{f_0(m)\}^{-1} x(x+1)\dots(x+m-1) \quad (29).$$

As to the complementary function it is apparent from (28) that we have

$$(\pi - a)^{-i}\, 0 = \rho^a \pi^{-i}\, 0.$$

Hence in particular if $i = 1$, we find

$$\begin{aligned}(\pi - a)^{-1}\, 0 &= \rho^a \pi^{-1}\, 0\\ &= \rho^a \Sigma x^{-1}\, 0\\ &= C\rho^a\\ &= Cx(x+1)\dots(x+a-1) \qquad (30).\end{aligned}$$

This method enables us to solve any equation of the form

$$x(x+1)\dots(x+n-1)\,\Delta^n u + A_1 x(x+1)\dots$$
$$\dots(x+n-2)\,\Delta^{n-1}u \dots + A_n u = X \dots\dots (31).$$

For symbolically expressed any such equation leads to the monomial form

$$\{\pi(\pi-1)\dots(\pi-n+1) + A_1\pi(\pi-1)\dots$$
$$\dots(\pi-n+2)\dots + A_n\}\, u = X \qquad (32).$$

Ex. 2. Given

$$x(x+1)\,\Delta^2 u - 2x\Delta u + 2u = x(x+1)(x+2).$$

The symbolical form of this equation is

$$\pi(\pi-1)\,u - 2\pi u + 2u = x(x+1)(x+2) \qquad (a),$$

or
$$(\pi^2 - 3\pi + 2)\, u = \rho^3.$$

Hence
$$\begin{aligned}u &= (\pi^2 - 3\pi + 2)^{-1}\rho^3\\ &= (3^2 - 3\times 3 + 2)^{-1}\rho^3\\ &\qquad + C_1\rho^2 + C_2\rho,\end{aligned}$$

since the factors of $\pi^2 - 3\pi + 2$ are $\pi - 2$ and $\pi - 1$. Thus we have

$$u = \frac{x(x+1)(x+2)}{2} + C_1 x(x+1) + C_2 x \qquad (b).$$

Binomial Equations.

7. Let us next suppose the given equation binomial and therefore susceptible of reduction to the form

$$u + \phi(\pi)\rho^n u = U \qquad (33),$$

in which U is a known, u the unknown and sought function of x. The possibility of finite solution will depend upon the form of the function $\phi(\pi)$, and its theory will consist of two parts, the first relating to the conditions under which the equation is directly resolvable into equations of the first order, the second to the laws of the transformations by which equations not obeying those conditions may when possible be reduced to equations obeying those conditions.

As to the first point it may be observed that if the equation be

$$u + \frac{1}{a\pi + b}\rho u = U \qquad (34),$$

it will, on reduction to the ordinary form, be integrable as an equation of the first order.

Again, if in (33) we have

$$\phi(\pi) = \psi(\pi)\psi(\pi - 1)\ldots\psi(\pi - n + 1),$$

in which $\psi(\pi) = \dfrac{1}{a\pi + b}$, the equation will be resolvable into a system of equations of the first order. This depends upon the general theorem that the equation

$$u + a_1\phi(\pi)\rho u + a_2\phi(\pi)\phi(\pi - 1)\rho^2 u \ldots$$
$$+ a_n\phi(\pi)\phi(\pi - 1)\ldots\phi(\pi - n + 1)\rho^n u = U$$

may be resolved into a system of equations, of the form

$$u - q\phi(\pi)\rho u = U,$$

q being a root of the equation

$$q^n + a_1 q^{n-1} + a_2 q^{n-2} \ldots + a_n = 0.$$

(*Differential Equations*, p. 405.)

Upon the same principle of formal analogy the propositions upon which the transformation of differential equations depends (*Ib.* pp. 408–9) might be adopted here with the mere substitution of π and ρ for E and ϵ^θ. But we prefer to investigate what may perhaps be considered as the most general forms of the theorems upon which these propositions rest.

From the binomial equation (33), expressed in the form

$$\{1 + \phi(\pi)\rho^n\} u = U,$$

we have

$$u = \{1 + \phi(\pi)\rho^n\}^{-1} U,$$

and this is a particular case of the more general form,

$$u = F\{\phi(\pi)\rho^n\} U \qquad (35).$$

Thus the unknown function u is to be determined from the known function U by the performance of a particular operation of which the *general* type is

$$F\{\phi(\pi)\rho^n\}.$$

Now suppose the given equations transformed by some process into a new but integrable binomial form,

$$v + \psi(\pi)\rho^n v = V,$$

V being here the given and v the sought function of x. We have

$$v = \{1 + \psi(\pi)\rho^n\}^{-1} V,$$

which is a particular case of $F\{\psi(\pi)\rho^n\} V$, supposing $F(t)$ to denote a function developable by Maclaurin's theorem. It is apparent therefore that the theory of this transformation must depend upon the theory of the connexion of the forms,

$$F\{\phi(\pi)\rho^n\}, \quad F\{\psi(\pi)\rho^n\}.$$

Let then the following inquiry be proposed. Given the forms of $\phi(\pi)$ and $\psi(\pi)$, is it possible to determine an operation $\chi(\pi)$ such that we shall have generally

$$F\{\phi(\pi)\rho^n\}\chi(\pi)X = \chi(\pi)F\{\psi(\pi)\rho^n\}X \qquad (36),$$

irrespectively of the form of X?

Supposing $F(t) = t$, we have to satisfy

$$\phi(\pi)\rho^n\chi(\pi)X = \chi(\pi)\psi(\pi)\rho^n X \qquad (37).$$

Hence by the first equation of (4),

$$\phi(\pi)\chi(\pi - n)\rho^n X = \psi(\pi)\chi(\pi)\rho^n X,$$

to satisfy which, independently of the form of X, we must have

$$\psi(\pi)\chi(\pi) = \phi(\pi)\chi(\pi - n);$$

$$\therefore \chi(\pi) = \frac{\phi(\pi)}{\psi(\pi)}\chi(\pi - n).$$

Therefore solving the above difference-equation,

$$\chi(\pi) = C\Pi_n\left\{\frac{\phi(\pi)}{\psi(\pi)}\right\}.$$

Substituting in (37), there results,

$$\phi(\pi)\rho^n\Pi_n\left\{\frac{\phi(\pi)}{\psi(\pi)}\right\}X = \Pi_n\left\{\frac{\phi(\pi)}{\psi(\pi)}\right\}\psi(\pi)\rho^n X,$$

or, replacing $\Pi_n\left\{\frac{\phi(\pi)}{\psi(\pi)}\right\}X$ by X_1,

and therefore X by $\left[\Pi_n\left\{\frac{\phi(\pi)}{\psi(\pi)}\right\}\right]^{-1}X_1$,

$$\phi(\pi)\rho^n X_1 = \Pi_n\left\{\frac{\phi(\pi)}{\psi(\pi)}\right\}\psi(\pi)\rho^n\left[\Pi_n\left\{\frac{\phi(\pi)}{\psi(\pi)}\right\}\right]^{-1}X_1.$$

If for brevity we represent $\Pi_n \left\{\frac{\phi(\pi)}{\psi(\pi)}\right\}$ by P, and drop the suffix from X_1 since the function is arbitrary, we have

$$\phi(\pi)\rho^n X = P\psi(\pi)\rho^n P^{-1}X.$$

Hence therefore

$$\{\phi(\pi)\rho^n\}^2 X = P\psi(\pi)\rho^n P^{-1}P\psi(\pi)\rho^n P^{-1}X$$
$$= P\{\psi(\pi)\rho^n\}^2 P^{-1}X,$$

and continuing the process,

$$\{\phi(\pi)\rho^n\}^m X = P\{\psi(\pi)\rho^n\}^m P^{-1}X.$$

Supposing therefore $F(t)$ to denote any function developable by Maclaurin's theorem, we have

$$F\{\phi(\pi)\rho^n\} X = PF\{\psi(\pi)\rho^n\} P^{-1}X.$$

We thus arrive at the following theorem.

THEOREM. *The symbols π and ρ combining in subjection to the law*

$$f(\pi)\rho^m X = \rho^m f(\pi + m) X,$$

the members of the following equation are symbolically equivalent, viz.

$$F\{\phi(\pi)\rho^n\} = \Pi_n \left\{\frac{\phi(\pi)}{\psi(\pi)}\right\} F\{\psi(\pi)\rho^n\} \Pi_n \left\{\frac{\psi(\pi)}{\phi(\pi)}\right\} \quad (38).$$

A. From this theorem it follows, in particular, that we can always convert the equation

$$u + \phi(\pi)\rho^n u = U$$

into any other binomial form,

$$v + \psi(\pi)\rho^n v = \Pi_n \left\{\frac{\psi(\pi)}{\phi(\pi)}\right\} U,$$

by assuming $u = \Pi_n \left\{\frac{\phi(\pi)}{\psi(\pi)}\right\} v$.

For we have

$$u = \{1 + \phi(\pi)\,\rho^n\}^{-1}\,U$$

$$= \Pi_n \left\{\frac{\phi(\pi)}{\psi(\pi)}\right\} \{1 + \psi(\pi)\,\rho^n\}^{-1}\, \Pi_n \left\{\frac{\psi(\pi)}{\phi(\pi)}\right\} U,$$

whence since

$$v = \{1 + \psi(\pi)\,\rho^n\}^{-1}\,V,$$

it follows that we must have

$$V = \Pi_n \left\{\frac{\psi(\pi)}{\phi(\pi)}\right\} U, \quad u = \Pi_n \left\{\frac{\phi(\pi)}{\psi(\pi)}\right\} v.$$

In applying the above theorem, it is of course necessary that the functions $\phi(\pi)$ and $\psi(\pi)$ be so related that the continued product denoted by $\Pi_n \left\{\frac{\phi(\pi)}{\psi(\pi)}\right\}$ should be finite. The conditions relating to the introduction of arbitrary constants have been stated with sufficient fulness elsewhere (*Differential Equations,* Chap. XVII. Art. 4).

B. The reader will easily demonstrate also the following theorem, viz.:

$$F\{\phi(\pi)\,\rho^n\}\,X = \rho^m F\{\phi(\pi + m)\,\rho^n\}\,\rho^{-m} X,$$

and deduce hence the consequence that the equation

$$u + \phi(\pi)\,\rho^n u = U$$

may be converted into

$$v + \phi(\pi + m)\,\rho^n v = \rho^{-m}\,U,$$

by assuming $u = \rho^m v$.

8. These theorems are in the following sections applied to the solution, or rather to the discovery of the conditions of finite solution, of certain classes of equations of considerable generality. In the first example the second member of the given equation is supposed to be any function of x. In the two others it is supposed to be 0. But the conditions of finite solution, if by this be meant the reduction of the discovery of the unknown quantity to the performance of a finite

number of operations of the kind denoted by Σ, will be the same in the one case as in the other. It is however to be observed, that when the second member is 0, a finite integral may be frequently obtained by the process for solutions in series developed in Art. 5, while if the second member be X, it is almost always necessary to have recourse to the transformations of Art. 7.

Discussion of the equation

$$(ax+b)\,u_x+(cx+e)\,u_{x-1}+(fx+g)\,u_{x-2}=X \qquad (a).$$

Consider first the equation

$$(ax+b)\,u_x+(cx+e)\,u_{x-1}+f(x-1)\,u_{x-2}=X \qquad (b).$$

Let $u_x=\mu^x v_x$, then, substituting, we have

$$\mu^2(ax+b)\,v_x+\mu(cx+e)\,v_{x-1}+f(x-1)\,v_{x-2}=\mu^{-x+2}X.$$

Multiply by x and assume $\pi=x\dfrac{\Delta}{\Delta x}$, $\rho=xE$, in which $\Delta x=-1$, then

$$\mu^2(ax^2+bx)\,v_x+\mu(cx+e)\,\rho v_x+f\rho^2 v_x=x\mu^{-x+2}X,$$

whence, substituting $\pi+\rho$ for x and developing the coefficients, we find

$$\mu^2(a\pi^2+b\pi)\,v_x+\mu\{(2a\mu+c)\,\pi+(b-a)\,\mu+e\}\,\rho v_x$$
$$+(a\mu^2+c\mu+f)\,\rho^2 v_x=x\mu^{-x+2}X \qquad (c),$$

and we shall now seek to determine μ so as to reduce this equation to a binomial form.

1st. Let μ be determined by the condition

$$a\mu^2+c\mu+f=0,$$

then making

$$2a\mu+c=A, \quad (b-a)\,\mu+e=B,$$

we have

$$a\mu\pi\left(\pi+\frac{b}{a}\right)v_x + A\left(\pi+\frac{B}{A}\right)\rho v_x = x\mu^{-x+1}X,$$

or

$$v_x + \frac{A}{a\mu}\frac{\pi+\frac{B}{A}}{\pi\left(\pi+\frac{b}{a}\right)}\rho v_x = \frac{1}{a\mu}\left\{\pi\left(\pi+\frac{b}{a}\right)\right\}^{-1} x\mu^{-x+1}X,$$

or, supposing V to be any *particular* value of the second member obtained by Art. 6, for it is not necessary at this stage to introduce an arbitrary constant,

$$v_x + \frac{A}{a\mu}\frac{\pi+\frac{B}{A}}{\pi\left(\pi+\frac{b}{a}\right)}\rho v_x = V \qquad (d).$$

This equation can be integrated when either of the functions,

$$\frac{B}{A},\ \frac{B}{A}-\frac{b}{a},$$

is an integer. In the former case we should assume

$$w_x + \frac{A}{a\mu}\frac{1}{\pi+\frac{b}{a}}\rho w_x = W_x \qquad (e),$$

whence we should have by (A),

$$v_x = \Pi_1\left(\frac{\pi+\frac{B}{A}}{\pi}\right)w_x,\quad W_x = \Pi_1\left(\frac{\pi}{\pi+\frac{B}{A}}\right)V \qquad (f).$$

In the latter case we should assume as the transformed equation

$$w_x + \frac{A}{a\mu}\frac{1}{\pi}\rho w_x = W_x \qquad (g),$$

and should find

$$v_x = \Pi_1 \left(\frac{\pi + \frac{B}{A}}{\pi + \frac{b}{a}} \right) w_x, \quad W_x = \Pi_1 \left(\frac{\pi + \frac{b}{a}}{\pi + \frac{B}{A}} \right) V \qquad (h).$$

The value of W_x obtained from (f) or (h) is to be substituted in (e) or (g), w_x then found by integration, and v_x determined by (f) or (h). One arbitrary constant will be introduced in the integration for w_x, and the other will be due either to the previous process for determining W_x, or to the subsequent one for determining v_x.

Thus in the particular case in which $\frac{B}{A}$ is a positive integer, we should have

$$W_x = \left\{ \left(\pi + \frac{B}{A}\right) \left(\pi + \frac{B}{A} - 1\right) \dots (\pi + 1) \right\}^{-1} 0,$$

a particular value of which, derived from the interpretation of $\left(\pi + \frac{B}{A}\right)^{-1} 0$ and involving an arbitrary constant, will be found to be $\frac{c}{1+x}$. Substituting in (e) and reducing the equation to the ordinary unsymbolical form, we have

$$\mu (ax + b) w_x + (A - \mu a) x w_{x-1} = \frac{c_1}{1+x},$$

and w_x being hence found, we have

$$v_x = \left(\pi + \frac{B}{A}\right) \left(\pi + \frac{B}{A} - 1\right) \dots (\pi + 1) w_x$$

for the complete integral.

2ndly. Let μ be determined so as if possible to cause the second term of (c) to vanish. This requires that we have

$$2a\mu + c = 0,$$

$$(b - a) \mu + e = 0,$$

and therefore imposes the condition

$$2ae + (b - a) c = 0.$$

Supposing this satisfied, we obtain, on making $\mu = \frac{-c}{2a}$,

$$v_x - \frac{h^2}{\pi\left(\pi + \frac{b}{a}\right)} \rho^2 v_x = \frac{1}{\mu^2 \pi \left(\pi + \frac{b}{a}\right)} x\mu^{-x+2} X,$$

or, representing any particular value of the second member by V,

$$v_x - \frac{h^2}{\pi\left(\pi + \frac{b}{a}\right)} \rho^2 v_x = V,$$

where

$$h = \frac{\sqrt{(c^2 - 4af)}}{c},$$

an equation which is integrable if $\frac{b}{a}$ be an odd number whether positive or negative. We must in such case assume

$$w_x - \frac{h^2}{\pi(\pi - 1)} \rho^2 w_x = W_x,$$

and determine first W_x and lastly v_x by h.

To found upon these results the conditions of solution of the general equation (a), viz.

$$(ax + b) u_x + (cx + e) u_{x-1} + (fx + g) u_{x-2} = X,$$

assume

$$fx + g = f(x' - 1),$$
$$u_x = t_{x'}.$$

Then

$$\left(ax' + b - a\frac{1+g}{f}\right) t_{x'}$$
$$+ \left(cx' + e - c\frac{1+f}{g}\right) t_{x'-1} + f(x' - 1) t_{x'-2} = X',$$

comparing which with (b) we see that it is only necessary in the expression of the conditions already deduced to change

$$b \text{ into } b - \frac{a(1+g)}{f}, \quad e \text{ into } e - \frac{c(1+g)}{f}.$$

Solution of the above equation when $X=0$ by definite integrals.*

9. If representing u_x by u we express (a) in the form

$$(ax+b)\,u+(cx+e)\,\epsilon^{-\frac{d}{dx}}u+(fx+g)\,\epsilon^{-2\frac{d}{dx}}u=0,$$

or

$$x\left(a+c\epsilon^{-\frac{d}{dx}}+f\epsilon^{-2\frac{d}{dx}}\right)u+\left(b+e\epsilon^{-\frac{d}{dx}}+g\epsilon^{-2\frac{d}{dx}}\right)u=0,$$

its solution in definite integrals may be obtained by Laplace's method for differential equations of the form

$$x\phi\left(\frac{d}{dx}\right)u+\psi\left(\frac{d}{dx}\right)u=0,$$

each particular integral of which is of the form

$$u=C\int\frac{\epsilon^{xt+\int\frac{\psi(t)}{\phi(t)}dt}}{\phi(t)}\,dt,$$

the limits of the final integration being any roots of the equation

$$\epsilon^{xt+\int\frac{\psi(t)}{\phi(t)}dt}=0.$$

See *Differential Equations*, Chap. XVIII.

The above solution is obtained by assuming $u=\int\epsilon^{xt}f(t)\,dt$, and then by substitution in the given equation and reduction obtaining a differential equation for determining the form of $f(t)$, and an algebraic equation for determining the limits. Laplace actually makes the assumption

$$u=\int t^x F(t)\,dt,$$

which differs from the above only in that $\log t$ takes the place of t and of course leads to equivalent results (*Théorie Analytique des Probabilités*, pp. 121, 135). And he employs this method with a view not so much to the solution of difficult equations as to the expression of solutions in forms convenient for calculation when functions of large numbers are involved.

* See also a paper by Thomæ (*Zeitschrift*, XIV. 349).

Thus taking his first example, viz.

$$u_{x+1} - (x+1)\, u_x = 0,$$

and assuming $u_x = \int t^x F(t)\, dt$, we have

$$\int t^{x+1} F(t)\, dt - (x+1) \int t^x F(t)\, dt = 0 \qquad (i).$$

But

$$(x+1) \int t^x F(t)\, dt = \int F(t)\, (x+1)\, t^x dt$$
$$= F(t)\, t^{x+1} - \int t^{x+1} F'(t)\, dt.$$

So that (i) becomes on substitution

$$\int t^{x+1} \{F(t) + F'(t)\}\, dt - F(t)\, t^{x+1} = 0,$$

and furnishes the two equations

$$F'(t) + F(t) = 0,$$
$$F(t)\, t^{x+1} = 0,$$

the first of which gives

$$F(t) = C\epsilon^{-t},$$

and thus reducing the second to the form

$$C\epsilon^{-t}\, t^{x+1} = 0,$$

gives for the limits $t = 0$ and $t = \infty$, on the assumption that $x+1$ is positive. Thus we have finally

$$u_x = C \int_0^\infty \epsilon^{-t}\, t^x dt,$$

the well-known expression for $\Gamma(x+1)$. A peculiar method of integration is then applied to convert the above definite integral into a rapidly convergent series.

Discussion of the equation

$$(ax^2 + bx + c)\, u_x + (ex + f)\, u_{x-1} + g u_{x-2} = 0 \qquad (a).$$

10. Let $u_x = \dfrac{\mu^x v_x}{1 \,.\, 2 \ldots x}$; then

$$\mu^2 (ax^2 + bx + c)\, v_x + \mu\, (ex + f)\, x v_{x-1} + g x\, (x-1)\, v_{x-2} = 0.$$

Whence, assuming $\pi = x\dfrac{\Delta}{\Delta x}$, $v = xE$, where $\Delta x = -1$, we have

$$\mu^2 (ax^2 + bx + c)\, v_x + \mu\, (ex + f)\, \rho v_x + g\rho^2 v_x = 0.$$

Therefore substituting $\pi + \rho$ for x, and developing by (13),

$$\mu^2 (a\pi^2 + b\pi + c) + \mu\, \{(2a\mu + e)\, \pi + (b - a)\, \mu + f\}\, \rho v_x$$
$$+ (\mu^2 a + \mu e + g)\, \rho^2 v_x = 0 \qquad (b).$$

First, let μ be determined so as to satisfy the equation

$$a\mu^2 + e\mu + g = 0,$$

then

$$\mu\, (a\pi^2 + b\pi + c)\, v_x + \{(2a\mu + e)\, \pi + (b - a)\, \mu + f\}\, \rho v_x = 0.$$

Whence, by Art. 5,

$$v_x = \Sigma a_m x\, (x - 1) \ldots (x - m + 1),$$

the successive values of a_m being determined by the equation

$$\mu^2 (am^2 + bm + c)\, a_m + \{(2a\mu^2 + e\mu)\, m + (b - a)\, \mu^2 + f\mu\}\, a_{m-1} = 0,$$

or
$$a_m = -\frac{(2a\mu + e)\, m + (b - a)\, \mu + f}{\mu\, (am^2 + bm + c)}\, a_{m-1}.$$

Represent this equation in the form

$$a_m = -f(m)\, a_{m-1},$$

and let the roots of the equation

$$am^2 + bm + c = 0$$

be α and β, then

$$v_x = C\, \{x^{(\alpha)} - f(\alpha + 1)\, x^{(\alpha+1)} + f(\alpha + 1)\, f(\alpha + 2)\, x^{(\alpha+2)} - \ldots\}$$
$$+ C'\, \{x^{(\beta)} - f(\beta + 1)\, x^{(\beta+1)} + f(\beta + 1)\, f(\beta + 2)\, x^{(\beta+2)} - \ldots\} \qquad (c),$$

where generally

$$x^{(p)} = x\, (x - 1) \ldots (x - p + 1).$$

One of these series will terminate whenever the value of m given by the equation

$$(2a\mu + e) m + (b - a) \mu + f = 0$$

exceeds by an integer either root of the equation

$$am^2 + bm + c = 0.$$

The solution may then be completed as in the last example.

Secondly, let μ be determined if possible so as to cause the second term of (b) to vanish. This gives

$$2a\mu + e = 0,$$

$$(b - a) \mu + f = 0,$$

whence, eliminating μ, we have the condition

$$2af + (a - b) e = 0.$$

This being satisfied, and μ being assumed equal to $-\dfrac{e}{2a}$, (b) becomes

$$(a\pi^2 + b\pi + c) v_x - \frac{a(e^2 - 4ag)}{e^2} \rho^2 v_x = 0.$$

Or putting
$$h = \frac{\sqrt{(e^2 - 4ag)}}{e},$$

$$v_x - \frac{h^2}{\pi^2 + \dfrac{b}{a}\pi + \dfrac{c}{a}} \rho^2 v_x = 0,$$

and is integrable in finite terms if the roots of the equation

$$m^2 + \frac{b}{a} m + \frac{c}{a} = 0$$

differ by an odd number.

Discussion of the equation

$$(ax^2 + bx + c) \Delta^2 u_x + (ex + f) \Delta u_x + g u_x = 0.$$

11. By resolution of its coefficients this equation is reducible to the form

$$a(x - \alpha)(x - \beta) \Delta^2 u_x + e(x - \gamma) \Delta u_x + g u_x = 0 \qquad (a).$$

Now let $x - \alpha = x' + 1$ and $u_x = v_{x'}$, then we have

$$a(x'+1)(x'+\alpha-\beta+1)\Delta^2 v_{x'} \\ + e(x'+\alpha-\gamma+1)\Delta v_{x'} + g v_{x'} = 0,$$

or, dropping the accent,

$$a(x+1)(x+\alpha-\beta+1)\Delta^2 v_x \\ + e(x+\alpha-\gamma+1)\Delta v_x + g v_x = 0 \quad (b).$$

If from the solution of this equation v_x be obtained, the value of u_x will thence be deduced by merely changing x into $x - \alpha - 1$.

Now multiply (b) by x, and assume

$$\pi = x\frac{\Delta}{\Delta x}\rho = xE,$$

where $\Delta x = 1$. Then, since by (20),

$$x(x+1)\Delta^2 v_x = \pi(\pi - 1) v_x,$$

we have

$$a(x+\alpha-\beta+1)\pi(\pi-1) v_x \\ + e(x+\alpha-\gamma+1)\pi v_x + g v_x = 0.$$

But $x = -\pi + \rho$, therefore substituting, and developing the coefficients we have on reduction

$$\pi\{a(\pi-\alpha+\beta-1)(\pi-1) + e(\pi-\alpha+\gamma-1) + g\} v_x \\ - \{a(\pi-1)(\pi-2) + e(\pi-1) + g\}\rho v_x = 0 \quad (c).$$

And this is a binomial equation whose solutions in series are of the form

$$v_x = \Sigma a_m x(x+1)\ldots(x+m-1),$$

the lowest value of m being a root of the equation

$$m\{a(m-\alpha+\beta-1)(m-1) + e(m-\alpha+\gamma-1) + g\} = 0 \quad (d),$$

corresponding to which value a_m is an arbitrary constant, while all succeeding values of a_m are determined by the law

$$a_m = \frac{a(m-1)(m-2) + e(m-1) + g}{m\{a(m-\alpha+\beta-1)(m-1) + e(m-\alpha+\gamma-1) + g\}} a_{m-1}.$$

Hence the series terminates when a root of the equation

$$a(m-1)(m-2)+e(m-1)+g=0 \qquad (e)$$

is equal to, or exceeds by an integer, a root of the equation (d).

As a particular root of the latter equation is 0, a particular finite solution may therefore always be obtained when (e) is satisfied either by a vanishing or by a positive integral value of m.

12. The general theorem expressed by (38) admits of the following generalization, viz.

$$F\{\pi, \phi(\pi)\rho^n\} = \Pi_n\left(\frac{\phi(\pi)}{\psi(\pi)}\right) F\{\pi, \psi(\pi)\rho^n\} \, \Pi_n\left(\frac{\psi(\pi)}{\phi(\pi)}\right).$$

The ground of this extension is that the symbol π, which is here newly introduced under F, combines with the same symbol π in the composition of the forms $\Pi_n\left(\frac{\phi(\pi)}{\psi(\pi)}\right)$, $\Pi_n\left(\frac{\psi(\pi)}{\phi(\pi)}\right)$ external to F, as if π were algebraic.

And this enables us to transform some classes of equations which are not binomial. Thus the solution of the equation

$$f_0(\pi)u + f_1(\pi)\phi(\pi)\rho u + f_2(\pi)\phi(\pi)\phi(\pi-1)\rho^2 u = U$$

will be made to depend upon that of the equation

$$f_0(\pi)v + f_1(\pi)\psi(\pi)\rho v + f_2(\pi)\psi(\pi)\psi(\pi-1)\rho^2 v = \Pi_1\left(\frac{\psi(\pi)}{\phi(\pi)}\right) U$$

by the assumption

$$u = \Pi_1\left(\frac{\phi(\pi)}{\psi(\pi)}\right) v.$$

13. While those transformations and reductions which depend upon the fundamental laws connecting π and ρ, and are expressed by (4), are common in their application to differential equations and to difference-equations, a marked difference exists between the two classes of equations as respects the conditions of finite solution. In differential equations where $\pi = \frac{d}{d\theta}$, $\rho = \epsilon^\theta$, there appear to be three primary integrable forms for binomial equations, viz.

$$u + \frac{a\pi + b}{c\pi + e}\rho^n u = U,$$

$$u + \frac{a(\pi - n)^2 + b}{\pi\left(\pi - \frac{n}{2}\right)} \rho^n u = U,$$

$$u + \frac{\left(\pi - \frac{n}{2}\right)(\pi - n)}{a\pi^2 + b} \rho^n u = U,$$

primary in the sense implied by the fact that every binomial equation, whatsoever its order, which admits of *finite* solution, is reducible to some one of the above forms by the transformations of Art. 7, founded upon the formal laws connecting π and ρ. In difference-equations but one primary integrable form for binomial equations is at present known, viz.

$$u + \frac{1}{a\pi + b} \rho u = U,$$

and this is but a particular case of the first of the above forms for differential equations. General considerations like these may serve to indicate the path of future inquiry.

14. Many attempts have been made to accomplish the general solution of linear difference-equations with variable coefficients, but the results are in all cases so complicated as to be practically useless. It will be sufficient if we mention Spitzer (*Grunert*, XXXII. and XXXIII.) on the class specially considered in this chapter, viz. when the coefficients are rational integral functions of the independent variable, Libri (*Crelle*, XII. 234), Binet (*Mémoires de l'Académie des Sciences*, XIX.). There is also a brief solution by Zehfuss (*Zeitschrift*, III. 177).

EXERCISES.

1. Of what theorem in the Differential Calculus does (20), Art. 4, constitute a generalization?

2. Solve the equation

$$x(x+1)\Delta^2 u + x\Delta u - n^2 u = 0.$$

3. Solve by the methods of Art. 7 the difference-equation of Ex. 1, Art. 5, supposing a to be a positive odd number.

4. Solve by the same methods the same equation, supposing a to be a negative odd number.

CHAPTER XIV.

MIXED AND PARTIAL DIFFERENCE-EQUATIONS.

1. If $u_{x,y}$ be any function of x and y, then

$$\left.\begin{aligned}\frac{\Delta}{\Delta x}u_{x,y}&=\frac{u_{x+\Delta x,y}-u_{x,y}}{\Delta x},\\ \frac{\Delta}{\Delta y}u_{x,y}&=\frac{u_{x,y+\Delta y}-u_{x,y}}{\Delta y},\end{aligned}\right\}\qquad(1).$$

These are, properly speaking, the coefficients of partial differences of the first order of $u_{x,y}$. But on the assumption that Δx and Δy are each equal to unity, an assumption which we can always legitimate, Chap. I. Art. 2, the above are the *partial differences* of the first order of $u_{x,y}$.

On the same assumption the general form of a partial difference of $u_{x,y}$ is

$$\frac{\Delta^{m+n}}{(\Delta x)^m(\Delta y)^n}u_{x,y}, \text{ or } \left(\frac{\Delta}{\Delta x}\right)^m\left(\frac{\Delta}{\Delta y}\right)^n u_{x,y} \qquad (2).$$

When the form of $u_{x,y}$ is given, this expression is to be interpreted by performing the successive operations indicated, each elementary operation being of the kind indicated in (1).

Thus we shall find

$$\frac{\Delta^2}{\Delta x\Delta y}u_{x,y}=u_{x+2\Delta x,\Delta y}-2u_{x+\Delta x,y+\Delta y}+u_{x,y+2\Delta y}.$$

It is evident that the operations $\frac{\Delta}{\Delta x}$ and $\frac{\Delta}{\Delta y}$ in combination are *commutative.*

Again, the symbolical expression of $\frac{\Delta}{\Delta x}$ in terms of $\frac{d}{dx}$ being

$$\frac{\Delta}{\Delta x} = \frac{\epsilon^{\Delta x \frac{d}{dx}} - 1}{\Delta x} \tag{3},$$

in which Δx is an absolute constant, it follows that

$$\left(\frac{\Delta}{\Delta x}\right)^n = \frac{\epsilon^{n\Delta x \frac{d}{dx}} - n\epsilon^{(n-1)\Delta x \frac{d}{dx}} + \frac{n(n-1)}{1.2}\epsilon^{(n-2)\Delta x \frac{d}{dx}} - \ldots}{(\Delta x)^n}$$

and therefore

$$\left(\frac{\Delta}{\Delta x}\right)^n u_{x,y} = \left\{u_{x+n\Delta x, y} - nu_{x+(n-1)\Delta x, y} + \frac{n(n-1)}{1.2} u_{x+(n-2)\Delta x, y} - \ldots\right\} \div (\Delta x)^n \tag{4}.$$

So, also, to express $\left(\frac{\Delta}{\Delta x}\right)^m \left(\frac{\Delta}{\Delta y}\right)^n u_{x,y}$ it would be necessary to substitute for $\frac{\Delta}{\Delta x}, \frac{\Delta}{\Delta y}$ their symbolical expressions, to effect their symbolical expansions by the binomial theorem, and then to perform the final operations on the subject function $u_{x,y}$.

Though in what follows each increment of an independent variable will be supposed equal to unity, it will still be necessary to retain the notation $\frac{\Delta}{\Delta x}, \frac{\Delta}{\Delta y}$ for the sake of distinction, or to substitute some notation equivalent by definition, e.g. Δ_x, Δ_y.

These things premised, we may define a partial difference-equation as an equation expressing an algebraic relation between any partial differences of a function $u_{x,y,z\ldots}$, the function itself, and the independent variables $x, y, z \ldots$ Or instead of the partial *differences* of the dependent function, its successive *values* corresponding to successive states of increment of the independent variables may be involved.

Thus
$$x \frac{\Delta}{\Delta x} u_{x,y} + y \frac{\Delta}{\Delta y} u_{x,y} = 0,$$

and
$$x u_{x+1,y} + y u_{x,y+1} - (x+y) u_{x,y} = 0,$$

are, on the hypothesis of Δx and Δy being each equal to unity, different but equivalent forms of the same partial difference-equation.

Mixed difference-equations are those in which the subject function is presented as modified both by operations of the form $\frac{\Delta}{\Delta x}$, $\frac{\Delta}{\Delta y}$, and by operations of the form $\frac{d}{dx}$, $\frac{d}{dy}$, singly or in succession. Thus

$$x \frac{\Delta}{\Delta x} u_{x,y} + y \frac{d}{dy} u_{x,y} = 0$$

is a mixed difference-equation. Upon the obvious subordinate distinction of ordinary mixed difference-equations and partial mixed difference-equations it is unnecessary to enter.

Partial Difference-equations.

2. When there are two independent variables x and y, while the coefficients are constant and the second member is 0, the proposed equation may be presented, according to convenience, in any of the forms

$$F(\Delta_x, \Delta_y)\, u = 0, \qquad F(E_x, E_y)\, u = 0,$$
$$F(\Delta_x, E_y)\, u = 0, \qquad F(E_x, \Delta_y)\, u = 0.$$

Now the symbol of operation relating to x, viz. Δ_x or E_x, combines with that relating to y, viz. Δ_y or E_y, as a constant with a constant. Hence a symbolical solution will be obtained by replacing one of the symbols by a constant quantity a, integrating the *ordinary* difference-equation which results, replacing a by the symbol in whose place it stands, and the arbitrary constant by an arbitrary function of the independent variable to which that symbol has reference. This arbitrary function must *follow* the expression which contains the symbol corresponding to a.

The condition last mentioned is founded upon the interpretation of $(E-a)^{-i}X$, upon which the solution of ordinary difference-equations with constant coefficients is ultimately dependent. For (Chap. XI. Art. 11)

$$(E-a)^{-i}X = a^{x-i}\Sigma^i a^{-x}X,$$

whence

$$(E-a)^{-i}0 = a^{x-i}\Sigma^i 0$$
$$= a^{x-i}(c_0 + c_1 x \ldots + c_{n-1}x^{n-1}),$$

the constants *following* the factor involving a.

The difficulty of the solution is thus reduced to the difficulty of interpreting the symbolical result.

Ex. 1. Thus the solution of the equation $u_{x+1} - au_x = 0$, of which the symbolical form is

$$E_x u_x - au_x = 0,$$

being

$$u_x = Ca^x,$$

the solution of the equation $u_{x+1,y} - u_{x,y+1} = 0$, of which the symbolic form is

$$E_x u_{x,y} - E_y u_{x,y} = 0,$$

will be

$$u_{x,y} = (E_y)^x \phi(y).$$

To interpret this we observe that since $E_y = \epsilon^{\frac{d}{dy}}$ we have

$$u_{x,y} = \epsilon^{x\frac{d}{dy}}\phi(y)$$
$$= \phi(y+x).$$

Ex. 2. Given $u_{x+1,y+1} - u_{x,y+1} - u_{x,y} = 0$.

This equation, on putting u for $u_{x,y}$, may be presented in the form

$$E_y \Delta_x u - u = 0. \qquad (1).$$

Now replacing E_y by a, the solution of the equation

$$a\Delta_x u - u = 0$$

is

$$u = (1 + a^{-1})^x C,$$

therefore the solution of (1) is

$$u = (1 + E_y^{-1})^x \phi(y) \qquad (2),$$

where $\phi(y)$ is an arbitrary function of y. Now, developing the binomial, and applying the theorem

$$E_y^{-n} \phi(y) = \phi(y - n),$$

we find

$$u = \phi(y) + x\phi(y-1) + \frac{x(x-1)}{1 \, . \, 2} \phi(y-2) + \dots \qquad (3),$$

which is finite when x is an integer.

Or, expressing (2) in the form

$$u = (E_y + 1)^x E_y^{-x} \phi(y),$$

developing the binomial in ascending powers of E_y, and interpreting, we have

$$u = \phi(y - x) + x\phi(y - x + 1)$$
$$+ \frac{x(x-1)}{1 \, . \, 2} \phi(y - x + 2) + \dots \qquad (4).$$

Or, treating the given equation as an ordinary difference-equation in which y is the independent variable, we find as the solution

$$u = (\Delta_x)^{-y} \phi(x) \qquad (5).$$

Any of these three forms may be used according to the requirements of the problem.

Thus if it were required that when $x = 0$, u should assume the form ϵ^{my}, it would be best to employ (3) or to revert to (2) which gives $\phi(y) = \epsilon^{my}$, whence

$$u = (1 + E_y^{-1})^x \epsilon^{my}$$
$$= (1 + \epsilon^{-\frac{d}{dy}})^x \epsilon^{my}$$
$$= (1 + \epsilon^{-m})^x \epsilon^{my} \qquad (6).$$

3. There is another method of integrating this class of equations with constant coefficients which deserves attention. We shall illustrate it by the last example.

Assume $u_{x,y} = \Sigma C a^x b^y$, then substituting in the given equation we find as the sole condition

$$ab - b - 1 = 0.$$

Hence

$$a = \frac{1+b}{b},$$

and substituting,

$$u_{x,y} = \Sigma C (1+b)^x b^{y-x}.$$

As the summation denoted by Σ has reference to all possible values of b, and C may vary in a perfectly arbitrary manner for different values of b, we shall best express the character of the solution by making C an arbitrary function of b and changing the summation into an integration extended from $-\infty$ to ∞. Thus we have

$$u_{x,y} = \int_{-\infty}^{\infty} b^{y-x} (1+b)^x \phi(b)\, db.$$

As $\phi(b)$ may be discontinuous, we may practically make the limits of integration what we please by supposing $\phi(b)$ to vanish when these limits are exceeded.

If we develope the binomial in ascending powers of b, we have

$$u_{x,y} = \int_{-\infty}^{\infty} b^{y-x} \phi(b)\, db + x \int_{-\infty}^{\infty} b^{y-x+1} \phi(b)\, db$$

$$+ \frac{x(x-1)}{1.2} \int_{-\infty}^{\infty} b^{y-x+2} \phi(b)\, db + \dots \qquad (7).$$

Now

$$\int_{-\infty}^{\infty} b^{\theta} \phi(b)\, db = \psi(\theta),$$

$\psi(\theta)$ being arbitrary if $\phi(b)$ is; hence

$$u_{x,y} = \psi(y-x) + x\psi(y-x+1) + \frac{x(x-1)}{1.2} \psi(y-x+2) + \dots,$$

which agrees with (4).

Although it is usually much the more convenient course to employ the symbolical method of Art. 2, yet cases may arise in which the expression of the solution by means of a definite integral will be attended with advantage; and the connexion of the methods is at least interesting.

Ex. 3. Given $\Delta^2_x u_{x-1,y} = \Delta^2_y u_{x,y-1}$.

Replacing $u_{x,y}$ by u, we have

$$(\Delta_x^2 E_x^{-1} - \Delta_y^2 E_y^{-1}) u = 0,$$

or $$(\Delta_x^2 E_y - \Delta_y^2 E_x) u = 0.$$

But $$\Delta_x = E_x - 1, \ \Delta_y = E_y - 1;$$

therefore $$(E_x^2 E_y + E_y - E_y^2 E_x - E_x) u = 0,$$

or $$(E_x E_y - 1)(E_x - E_y) u = 0.$$

This is resolvable into the two equations

$$(E_x E_y - 1) u = 0, \quad (E_x - E_y) u = 0.$$

The first gives

$$E_x u - E_y^{-1} u = 0,$$

of which the solution is

$$u = (E_y^{-1})^x \phi(y)$$
$$= \phi(y - x).$$

The second gives, by Ex. 1,

$$u = \psi(x + y).$$

Hence the complete integral is

$$u = \phi(y - x) + \psi(y + x).$$

4. Upon the result of this example an argument has been founded for the discontinuity of the arbitrary functions which occur in the solution of the partial differential equation

$$\frac{d^2u}{dx^2} - \frac{d^2u}{dy^2} = 0,$$

and thence, by obvious transformation, in that of the equation

$$\frac{d^2u}{dx^2} - a^2 \frac{d^2u}{dt^2} = 0.$$

It is perhaps needless for me, after what has been said in Chap. X., to add that I regard the *argument* as unsound. Analytically such questions depend upon the following, viz. whether in the proper sense of the term limit, we can regard $\sin x$ and $\cos x$ as tending to the limit 0, when x tends to become infinite.

5. When together with Δ_x and Δ_y one only of the independent variables, e.g. x, is involved, or when the equation contains both the independent variables, but only one of the operative symbols Δ_x, Δ_y, the same principle of solution is applicable. A symbolic solution of the equation

$$F(x, \Delta_x, \Delta_y)\, u = 0$$

will be found by substituting Δ_y for a and converting the arbitrary constant into an arbitrary function of y in the solution of the ordinary equation

$$F(x, \Delta_x, a)\, u = 0.$$

And a solution of the equation

$$F(x, y, \Delta_x) = 0$$

will be obtained by integrating as if y were a constant, and replacing the arbitrary constant, as before, by an arbitrary function of y. But if x, y, Δ_x and Δ_y are involved together, this principle is no longer applicable. For although y and Δ_y are constant relatively to x and Δ_x, they are not so with respect to each other. In such cases we must endeavour by a change of variables, or by some tentative hypothesis as to the form of the solution, to reduce the problem to easier conditions.

The extension of the method to the case in which the second member is not equal to 0 involves no difficulty.

Ex. 4. Given $\quad u_{x,y} - xu_{x-1,y-1} = 0.$

Writing u for $u_{x,y}$ the equation may be expressed in the form

$$u - xE_x^{-1}E_y^{-1}u = 0 \qquad (1).$$

Now replacing E_y^{-1} by a, the solution of

$$u - axE_x^{-1}u = 0 \text{ or } u_x - axu_{x-1} = 0$$

is

$$Cx(x-1)\ldots 1 \,.\, a^x.$$

Wherefore, changing a into E_y^{-1}, the solution of (1) is

$$\begin{aligned} u &= (E_y^{-1})^x x(x-1)\ldots 1 \,.\, \phi(y) \\ &= x(x-1)\ldots 1 \,.\, (E_y^{-x})\,\phi(y) \\ &= x(x-1)\ldots 1 \,.\, \phi(y-x). \end{aligned}$$

6. Laplace has shewn how to solve any linear equation in the successive terms of which the progression of differences is the same with respect to one independent variable as with respect to the other.

The given equation being

$$A_{x,y}u_{x,y} + B_{x,y}u_{x-1,y-1} + C_{x,y}u_{x-2,y-2} + \ldots = V_{x,y},$$

$A_{x,y}$, $B_{x,y}$, …, being functions of x and y, let $y = x - k$;

then substituting and representing $u_{x,y}$ by v_x, the equation assumes the form

$$X_0v_x + X_1v_{x-1} + X_2v_{x-2} + \ldots = X,$$

X_0, $X_1 \ldots X$ being functions of x. This being integrated, k is replaced by $x - y$, and the arbitrary constants by arbitrary functions of $x - y$.

The ground of this method is that the progression of differences in the given equation is such as to leave $x - y$ unaffected, for when x and y change by equal differences $x - y$ is unchanged. Hence if $x - y$ is represented by k and we take x and k for the new variables, the differences now having reference to x only, we can integrate as if k were constant.

Applying this method to the last example, we have

$$v_x - xv_{x-1} = 0,$$

$$v_x = cx(x-1)\ldots 1,$$

$$u_{x,y} = x(x-1)\ldots 1 \,.\, \phi(x-y),$$

which agrees with the previous result.

The method may be generalized. Should any linear function of x and y, e.g. $x+y$, be invariable, we may by assuming it as one of the independent variables, so to speak reduce the equation to an ordinary difference-equation; but arbitrary functions of the element in question must take the place of arbitrary constants.

Ex. 5. Given $u_{x,y} - pu_{x+1,y-1} - (1-p)\,u_{x-1,y+1} = 0$.

Here $x+y$ is invariable. Now the integral of

$$v_x - pv_{x+1} - (1-p)\,v_{x-1} = 0$$

is
$$v_x = c + c'\left(\frac{1-p}{p}\right)^x.$$

Hence, that of the given equation is

$$u_{x,y} = \phi(x+y) + \left(\frac{1-p}{p}\right)^x \psi(x+y).$$

7. Partial difference-equations are of frequent occurrence in the theory of games of chance. The following is an example of the kind of problems in which they present themselves.

Ex. 6. A and B engage in a game, each step of which consists in one of them winning a counter from the other. At the commencement, A has x counters and B has y counters, and in each successive step the probability of A's winning a counter from B is p, and therefore of B's winning a counter from A, $1-p$. The game is to terminate when either of the two has n counters. What is the probability of A's winning it?

Let $u_{x,y}$ be the probability that A will win it, any positive values being assigned to x and y.

Now A's winning the game may be resolved into two alternatives, viz. 1st, His winning the first step, and afterwards winning the game. 2ndly, His losing the first step, and afterwards winning the game.

The probability of the first alternative is $pu_{x+1, y-1}$, for after A's winning the first step, the probability of which is p, he will have $x+1$ counters, B, $y-1$ counters, therefore the probability that A will then win is $u_{x+1, y-1}$. Hence the probability of the combination is $pu_{x+1, y-1}$.

The probability of the second alternative is in like manner $(1-p)\, u_{x-1, y+1}$.

Hence, the probability of any event being the sum of the probabilities of the alternatives of which it is composed, we have as the equation of the problem

$$u_{x,y} = pu_{x+1, y-1} + (1-p)\, u_{x-1, y+1} \qquad (1),$$

the solution of which is, by the last example,

$$u_{x,y} = \phi(x+y) + \left(\frac{1-p}{p}\right)^x \psi(x+y).$$

It remains to determine the arbitrary functions.

The number of counters $x+y$ is invariable through the game. Represent it by m, then

$$u_{x,y} = \phi(m) + \left(\frac{1-p}{p}\right)^x \psi(m).$$

Now A's success is certain if he should ever be in possession of n counters. Hence, if $x = n$, $u_{x,y} = 1$. Therefore

$$1 = \phi(m) + \left(\frac{1-p}{p}\right)^n \psi(m).$$

Again, A loses the game if ever he have only $m-n$ counters, since then B will have n counters. Hence

$$0 = \phi(m) + \left(\frac{1-p}{p}\right)^{m-n} \psi(m).$$

The last two equations give, on putting $P=\dfrac{1-p}{p}$,

$$\phi(m)=\frac{-P^{m-n}}{P^n-P^{m-n}},\ \psi(m)=\frac{1}{P^n-P^{m-n}},$$

whence

$$u_{x,y}=\frac{P^{n-y}-1}{P^{2n-x-y}-1}$$

$$=\frac{\{p^{n-y}-(1-p)^{n-y}\}\,p^{n-x}}{p^{2n-x-y}-(1-p)^{2n-x-y}} \qquad (2),$$

which is the probability that A will win the game.

Symmetry therefore shews that the probability that B will win the game is

$$\frac{\{(1-p)^{n-x}-p^{n-x}\}\,p^{n-y}}{(1-p)^{2n-x-y}-p^{2n-x-y}} \qquad (3),$$

and the sum of these values will be found to be unity.

The problem of the 'duration of play' in which it is proposed to find the probability that the game conditioned as above will terminate at a particular step, suppose the rth, depends on the same partial difference-equation, but it involves great difficulty. A very complete solution, rich in its analytical consequences, will be found in a memoir by the late Mr Leslie Ellis (*Cambridge Mathematical Journal*, Vol. IV. p. 182).

Method of Generating Functions.

8. Laplace usually solves problems of the above class by the method of generating functions, the most complete statement of which is contained in the following theorem.

Let u be the generating function of $u_{m,n\ldots}$, so that

$$u=\Sigma u_{m,n\ldots}x^m y^n\ldots,$$

then making $x=\epsilon^\theta$, $y=\epsilon^{\theta'}$, ... we have

$$S\phi\left(\frac{d}{d\theta},\ \frac{d}{d\theta'},\ \ldots\right)\epsilon^{p\theta+q\theta'\ldots}u$$

$$=\Sigma\{S\phi(m,n\ldots)\,u_{m-p,\,n-q\ldots}\}\,\epsilon^{m\theta+n\theta'\ldots} \qquad (1).$$

Here, while Σ denotes summation with respect to the terms of the development of u, S denotes summation with respect to the operations which would constitute the first member a member of a linear differential equation, and the bracketed portion of the second member a member of a difference-equation.

Hence it follows that if we have a linear difference-equation of the form

$$S\phi\,(m, n \ldots)\, u_{m-p,\, n-q\ldots} = 0 \qquad (2),$$

the equation (1) would give for the general determination of the generating function u the linear differential equation

$$\Sigma\phi\left(\frac{d}{d\theta}, \frac{d}{d\theta'} \ldots\right) \epsilon^{p\theta+q\theta'\ldots}\, u = 0 \qquad (3).$$

But if there be given certain initial values of $u_{m,n}$ which the difference-equation does not determine, then, corresponding to such initial values, terms will arise in the second member of (1) so that the differential equation will assume the form

$$S\phi\left(\frac{d}{d\theta}, \frac{d}{d\theta'} \ldots\right) \epsilon^{p\theta+q\theta'\ldots}\, u = F\,(m, n, \ldots) \qquad (4).$$

If the difference-equation have constant coefficients the differential equation merges into an algebraic one, and the generating function will be a rational fraction. This is the case in most, if not all, of Laplace's examples.

It must be borne in mind that the discovery of the generating function is but a step toward the solution of the difference-equation, and that the next step, viz. the discovery of the general term of its development by some *independent* process, is usually far more difficult than the direct solution of the original difference-equation would be. As I think that in the present state of analysis the interest which belongs to this application of generating functions is chiefly historical, I refrain from adding examples.

Mixed Difference-equations.

9. When a mixed difference-equation admits of resolution into a simple difference-equation and a differential equation, the process of solution is obvious.

Ex. 7. Thus the equation

$$\Delta \frac{du}{dx} - a\Delta u - b\frac{du}{dx} + abu = 0$$

being presented in the form

$$\left(\frac{d}{dx} - a\right)(\Delta - b)\, u = 0,$$

the complete value of u will evidently be the sum of the values given by the resolved equations

$$\frac{du}{dx} - au = 0, \quad \Delta u - bu = 0.$$

Hence

$$u = c_1 \epsilon^{ax} + c_2 (1 + b)^x,$$

where c_1 is an absolute, c_2 a periodical constant.

Ex. 8. Again, the equation

$$\Delta y = x\frac{d}{dx}\Delta y + \left(\frac{d}{dx}\Delta y\right)^2$$

being resolvable into the two equations,

$$\Delta y = z, \quad z = x\frac{dz}{dx} + \left(\frac{dz}{dx}\right)^2,$$

we have, on integration,

$$z = cx + c^2,$$

$$y = \Sigma z = \frac{cx(x-1)}{2} + c^2 x + C,$$

where c is an absolute, and C a periodical constant.

Mixed difference-equations are reducible to differential equations of an exponential form by substituting for E_x or Δ_x their differential expressions $\epsilon^{\frac{d}{dx}}$, $\epsilon^{\frac{d}{dx}} - 1$.

Ex. 9. Thus the equation $\Delta u - \frac{du}{dy} = 0$ becomes

$$\left(\epsilon^{\frac{d}{dy}} - 1 - \frac{d}{dy}\right) u = 0,$$

and its solution will therefore be

$$u = \Sigma c_m \epsilon^{my},$$

the values of m being the different roots of the equation

$$\epsilon^m - 1 - m = 0.$$

10. Laplace's method for the solution of a class of partial differential equations (*Diff. Equations*, p. 440) has been extended by Poisson to the solution of mixed difference-equations of the form

$$\frac{du_{x+1}}{dx} + L\frac{du_x}{dx} + Mu_{x+1} + Nu_x = V \qquad (1),$$

where L, M, N, V are functions of x.

Writing u for u_x, and expressing the above equation in the form

$$\frac{d}{dx}Eu + L\frac{d}{dx}u + MEu + Nu = V,$$

it is easily shewn that it is reducible to the form

$$\left(\frac{d}{dx} + M\right)(E + L)\,u + (N - LM - L')\,u = V,$$

where $L' = \frac{dL}{dx}$. Hence if we have

$$N - LM - L' = 0 \qquad (2),$$

the equation becomes

$$\left(\frac{d}{dx}+M\right)(E+L)\,u=V,$$

which is resolvable by the last section into a mixed difference-equation and a differential equation.

But if the above condition be not satisfied, then, assuming

$$(E+L)\,u=v \qquad (3),$$

we have

$$\left(\frac{d}{dx}+M\right)v+(N-LM-L')\,u=V,$$

whence

$$u=\frac{-\left(\frac{d}{dx}+M\right)v+V}{N-LM-L'} \qquad (4),$$

which is expressible in the form

$$u=A_x\frac{dv}{dx}+B_xv+C_x.$$

Substituting this value in (3) we have

$$A_{x+1}\frac{d}{dx}Ev+LA_x\frac{dv}{dx}+B_{x+1}Ev$$

$$+(LB_x-1)\,v_x=-C_{x+1}-LC_x,$$

which, on division by A_{x+1}, is of the form

$$\frac{d}{dx}Ev+L_1\frac{dv}{dx}+M_1Ev+N_1v=V_1.$$

The original form of the equation is thus reproduced with altered coefficients, and the equation is resolvable as before into a mixed difference-equation and a differential equation, if the condition

$$N_1-L_1M_1-L_1'=0 \qquad (5)$$

is satisfied. If not, the operation is to be repeated.

An inversion of the order in which the symbols $\frac{d}{dx}$ and E are employed in the above process leads to another reduction similar in its general character.

Presenting the equation in the form

$$(E+L)\left(\frac{d}{dx}+M_{-1}\right)u+(N-LM_{-1})\,u=V$$

where $M_{-1}=E^{-1}M$, its direct resolution into a mixed difference-equation and a differential equation is seen to involve the condition

$$N-LM_{-1}=0. \qquad (6).$$

If this equation be not satisfied, assume

$$\left(\frac{d}{dx}+M_{-1}\right)u=v,$$

and proceeding as before a new equation similar in form to the original one will be obtained to which a similar test, or, that test failing, a similar reduction may again be applied.

Ex. 10. Given $\frac{du_{x+1}}{dx}-a\frac{du_x}{dx}+(x\pm n)\,u_{x+1}-axu_x=0.$

This is the most general of Poisson's examples. Taking first the lower sign we have

$$L=-a,\ M=x-n,\ N=-ax.$$

Hence the condition (2) is not satisfied. But (3) and (4) give

$$(E-a)\,u=v,$$

$$u=\frac{\frac{dv}{dx}+(x-n)\,v}{an},$$

whence

$$(E-a)\left\{\frac{\frac{dv}{dx}+(x-n)\,v}{an}\right\}=v,$$

or, on reducing,

$$\frac{dv_{x+1}}{dx} - a\frac{dv_x}{dx} + \{x - (n-1)\} v_{x+1} - axv_x = 0.$$

Comparing this with the given equation, we see that n reductions similar to the above will result in an equation of the form

$$\frac{dw_{x+1}}{dx} - a\frac{dw_x}{dx} + xw_{x+1} - axw_x = 0,$$

which, being presented in the form

$$\left(\frac{d}{dx} + x\right)(E - a)\, w_x = 0,$$

is resolvable into two equations of the unmixed character.

Poisson's second reduction applies when the upper sign is taken in the equation given; and thus the equation is seen to be integrable whenever n is an integer positive or negative.

Its actual solution deduced by another method will be given in the following section.

11. Mixed difference-equations in whose coefficients x is involved only in the first degree admit of a symbolical solution founded upon the theorem

$$\left\{x + \phi'\left(\frac{d}{dx}\right)\right\}^{-1} X = \epsilon^{\phi\left(\frac{d}{dx}\right)} x^{-1} \epsilon^{-\phi\left(\frac{d}{dx}\right)} X \qquad (1).$$

(*Differential Equations*, p. 445.)

The following is the simplest proof of the above theorem. Since

$$\psi\left(\frac{d}{dx}\right) xu = \psi\left(\frac{d}{dx} + \frac{d'}{dx}\right) xu,$$

if in the second member $\frac{d'}{dx}$ operate on x only, and $\frac{d}{dx}$ on u, we have, on developing and effecting the differentiations which have reference to x,

$$\psi\left(\frac{d}{dx}\right) xu = x\psi\left(\frac{d}{dx}\right) u + \psi'\left(\frac{d}{dx}\right) u.$$

Let $\psi\left(\frac{d}{dx}\right) u = v$, then

$$\psi\left(\frac{d}{dx}\right) x \left\{\psi\left(\frac{d}{dx}\right)\right\}^{-1} v = \left\{x + \frac{\psi'\left(\frac{d}{dx}\right)}{\psi\left(\frac{d}{dx}\right)}\right\} v,$$

or if $\psi\left(\frac{d}{dx}\right)$ be replaced by $\epsilon^{\phi\left(\frac{d}{dx}\right)}$,

$$\epsilon^{\phi\left(\frac{d}{dx}\right)} x \epsilon^{-\phi\left(\frac{d}{dx}\right)} v = \left\{x + \phi'\left(\frac{d}{dx}\right)\right\} v.$$

Inverting the operations on both sides, which involves the inverting of the order as well as of the character of successive operations, we have

$$\left\{x + \phi'\left(\frac{d}{dx}\right)\right\}^{-1} v = \epsilon^{\phi\left(\frac{d}{dx}\right)} x^{-1} \epsilon^{-\phi\left(\frac{d}{dx}\right)} v,$$

the theorem in question.

Let us resume Ex. 10, which we shall express in the form

$$\frac{du_{x+1}}{dx} - a\frac{du_x}{dx} + (x+n) u_{x+1} - axu_x = 0 \qquad (a),$$

n being either positive or negative. Now putting u for u_x

$$\left\{\frac{d}{dx}\left(\epsilon^{\frac{d}{dx}} - a\right) + n\epsilon^{\frac{d}{dx}}\right\} u + x\left(\epsilon^{\frac{d}{dx}} - a\right) u = 0.$$

Let $$\left(\epsilon^{\frac{d}{dx}} - a\right) u = z,$$

then we have

$$\left\{\frac{d}{dx} + n\epsilon^{\frac{d}{dx}}\left(\epsilon^{\frac{d}{dx}} - a\right)^{-1}\right\} z + xz = 0.$$

Or,

$$\left(x + \frac{d}{dx} + \frac{n\epsilon^{\frac{d}{dx}}}{\epsilon^{\frac{d}{dx}} - a}\right) z = 0.$$

Hence,

$$z = \left(x + \frac{d}{dx} + \frac{n\epsilon^{\frac{d}{dx}}}{\epsilon^{\frac{d}{dx}} - a}\right)^{-1} 0,$$

and therefore by (1),

$$z = \epsilon^{\frac{1}{2}\left(\frac{d}{dx}\right)^2 + n\log\left(\epsilon^{\frac{d}{dx}} - a\right)} x^{-1} \epsilon^{-\frac{1}{2}\left(\frac{d}{dx}\right)^2 - n\log\left(\epsilon^{\frac{d}{dx}} - a\right)} 0$$

$$= (\epsilon^{\frac{d}{dx}} - a)^n \epsilon^{\frac{1}{2}\left(\frac{d}{dx}\right)^2} x^{-1} \epsilon^{-\frac{1}{2}\left(\frac{d}{dx}\right)^2} (\epsilon^{\frac{d}{dx}} - a)^{-n} 0 \qquad (b).$$

It is desirable to transform a part of this expression. By (1), we have

$$\left(x + \frac{d}{dx}\right)^{-1} = \epsilon^{\frac{1}{2}\left(\frac{d}{dx}\right)^2} x^{-1} \epsilon^{-\frac{1}{2}\left(\frac{d}{dx}\right)^2},$$

and by another known theorem,

$$\left(\frac{d}{dx} + x\right)^{-1} = \epsilon^{-\frac{1}{2}x^2} \left(\frac{d}{dx}\right)^{-1} \epsilon^{\frac{1}{2}x^2}.$$

The right-hand members of these equations being symbolically equivalent, we may therefore give to (b) the form

$$z = (\epsilon^{\frac{d}{dx}} - a)^n \epsilon^{-\frac{x^2}{2}} \left(\frac{d}{dx}\right)^{-1} \epsilon^{\frac{x^2}{2}} (\epsilon^{\frac{d}{dx}} - a)^{-n} 0. \qquad (c).$$

Now $u = (\epsilon^{\frac{d}{dx}} - a)^{-1} z$, therefore substituting, and replacing $\epsilon^{\frac{d}{dx}}$ by E,

$$u = (E - a)^{n-1} \epsilon^{-\frac{x^2}{2}} \left(\frac{d}{dx}\right)^{-1} \epsilon^{\frac{x^2}{2}} (E - a)^{-n} 0 \qquad (A).$$

Two cases here present themselves.

First, let n be a positive integer; then since

$$(E-a)^{-n}0=a^x(c_0+c_1x\ldots+c_{n-1}x^{n-1}),$$

$$(E-a)^{n-1}=(\Delta+1-a)^{n-1},$$

we have

$$u=(\Delta+1-a)^{n-1}\epsilon^{-\frac{x^2}{2}}\{C+\int\epsilon^{\frac{x^2}{2}}a^x(c_0+c_1x\ldots+c_{n-1}x^{n-1})\,dx\} \qquad (d),$$

as the solution required.

This solution involves superfluous constants. For integrating by parts, we have

$$\int\epsilon^{\frac{x^2}{2}}a^xx^rdx=\epsilon^{\frac{x^2}{2}}a^xx^{r-1}+\log a\int\epsilon^{\frac{x^2}{2}}a^xx^{r-1}dx+(r-1)\int\epsilon^{\frac{x^2}{2}}a^xx^{r-2}dx,$$

and in particular when $r=1$,

$$\int\epsilon^{\frac{x^2}{2}}a^xxdx=\epsilon^{\frac{x^2}{2}}a^x+\log a\int\epsilon^{\frac{x^2}{2}}a^xdx.$$

These theorems enable us, r being a positive integer, to reduce the above general integral to a linear function of the elementary integrals $\int\epsilon^{\frac{x^2}{2}}a^xdx$, and of certain algebraic terms of the form $\epsilon^{\frac{x^2}{2}}a^xx^m$, where m is an integer less than r.

Now if we thus reduce the integrals involved in (d), it will be found that the algebraic terms vanish.

For

$$\begin{aligned}(\Delta+1-a)^{n-1}\epsilon^{-\frac{x^2}{2}}(\epsilon^{\frac{x^2}{2}}a^xx^m)&=(\Delta+1-a)^{n-1}a^xx^m\\&=a^{x+n-1}\Delta^{n-1}x^m\\&=0,\end{aligned}$$

since m is less than r, and the greatest value of r is $n-1$.

It results therefore that (d) assumes the simpler form,

$$u=(\Delta+1-a)^{n-1}\epsilon^{\frac{-x^2}{2}}(C_0+C_1\int\epsilon^{\frac{x^2}{2}}a^xdx);$$

and here C_0 introduced by ordinary integration is an absolute constant, while C_1 introduced by the performance of the operation Σ is a periodical constant.

A superfluity among the arbitrary constants, but a superfluity which does not affect their arbitrariness, is always to be *presumed* when the inverse operations by which they are introduced are at a subsequent stage of the process of solution followed by the corresponding *direct* operations. The particular observations of Chap. XVII. Art. 4 (*Differential Equations*) on this subject admit of a wider application.

Secondly, let n be 0 or a negative integer.

It is here desirable to change the sign of n so as to express the given equation in the form

$$\frac{du_1}{dx} - a\frac{du}{dx} + (x-n)\,u_1 - axu = 0,$$

while its symbolical solution (A) becomes

$$u = (E-a)^{-n-1}\,\epsilon^{\frac{-x^2}{2}}\left(\frac{d}{dx}\right)^{-1}\epsilon^{\frac{x^2}{2}}\,(E-a)^n\,0.$$

And in both n is 0 or a positive integer.

Now since $(E-a)^n\,0 = 0$, and $\left(\frac{d}{dx}\right)^{-1} 0 = C$, we have

$$\begin{aligned} u &= (E-a)^{-n-1}\,C\epsilon^{\frac{-x^2}{2}} \\ &= C(E-a)^{-n-1}\,\epsilon^{\frac{-x^2}{2}} + (E-a)^{-n-1}\,0 \\ &= Ca^{x-n-1}\,\Sigma^{n+1}\,a^{-x}\,\epsilon^{\frac{-x^2}{2}} + a^{x-n-1}\,\Sigma^{n+1}\,0 \\ &= C_1 a^x \Sigma^{n+1} a^{-x}\epsilon^{\frac{-x^2}{2}} + a^x\,(c_0 + c_1 x \ldots + c_n x^n). \end{aligned}$$

But here, while the absolute constant C_1 is arbitrary, the $n+1$ periodical constants $c_0, c_1, \ldots, c_n$ are connected by n relations which must be determined by substitution of the above unreduced value of u in the given equation.

The general expression of these relations is somewhat complex; but in any particular case they may be determined without difficulty.

Thus if $a = 1$, $n = 1$, it will be found that

$$u = C_1 \Sigma^2 a^{-x} \epsilon^{\frac{-x^2}{2}} + C_2 (1 - x).$$

If $a = 1$, $n = 2$, we shall have

$$u = C_1 \Sigma^3 a^{-x} \epsilon^{-\frac{x^2}{2}} + C_2 \left(1 - x + \frac{x^2}{3}\right),$$

and so on.

The two general solutions may be verified, though not easily, by substitution in the original equation.

12. The same principles of solution are applicable to mixed partial difference-equations as to partial difference-equations. If Δ_x and $\frac{d}{dy}$ are the symbols of pure operation involved, and if, replacing one of these by a constant m, the equation becomes either a pure differential equation or a pure difference-equation with respect to the other, then it is only necessary to replace in the solution of that equation m by the symbol for which it stands, to effect the corresponding change in the arbitrary constant, and then to interpret the result.

Ex. 11. $\Delta_x u - a\frac{du}{dy} = 0.$

Replacing $\frac{d}{dy}$ by m, and integrating, we have

$$u = c\,(1 + am)^x.$$

Hence the symbolic solution of the given equation is

$$u = \left(1 + a\frac{d}{dy}\right)^x \phi(y)$$

$$= a^x \left(\frac{d}{dy} + \frac{1}{a}\right)^x \phi(y)$$

$$= a^x \epsilon^{-\frac{y}{a}} \left(\frac{d}{dy}\right)^x \epsilon^{\frac{y}{a}} \phi(y),$$

$$= a^x \epsilon^{-\frac{y}{a}} \left(\frac{d}{dy}\right)^x \psi(y),$$

$\psi(y)$ being an arbitrary function of y.

Ex. 12. Given $u_{x+1,y} - \dfrac{d}{dy} u_{x,y} = V_{x,y}$.

Treating $\dfrac{d}{dy}$ as a constant, the symbolic solution is

$$u_{x,y} = \left(\frac{d}{dy}\right)^{x-1} \Sigma \left(\frac{d}{dy}\right)^{-x} V_{x,y} + \left(\frac{d}{dy}\right)^x \phi(y),$$

Σ having reference to x. No constants need to be introduced in performing the integrations implied by $\left(\dfrac{d}{dy}\right)^{-x}$.

Ex. 13. Given $u_{x+2} - 3x \dfrac{du_{x+1}}{dy} + 2x(x-1) \dfrac{d^2u_x}{dy^2} = 0$.

Let $u_x = 1 . 2 \ldots (x-2)\, v_x$, then

$$v_{x+2} - 3 \frac{dv_{x+1}}{dy} + 2 \frac{d^2v_x}{dy^2} = 0,$$

or

$$\left\{E_x^2 - 3E_x \frac{d}{dy} + 2\left(\frac{d}{dy}\right)^2\right\} v_x = 0,$$

or

$$\left(E_x - \frac{d}{dy}\right)\left(E_x - 2\frac{d}{dy}\right) v_x = 0,$$

whence by resolution and integration

$$v_x = \left(\frac{d}{dy}\right)^x \phi(y) + \left(2\frac{d}{dy}\right)^x \psi(y),$$

$$u_x = 1 . 2 \ldots (x-2) \left\{\left(\frac{d}{dy}\right)^x \phi(y) + 2^x \left(\frac{d}{dy}\right)^x \psi(y)\right\}.$$

Ex. 14. $u_{x+2} - 3 \dfrac{du_{x+1}}{dy} + 2 \dfrac{d^2u_x}{dy^2} = V$, where V is a function of x and y.

Here we have

$$\left\{E_x^2 - 3E_x\frac{d}{dy} + 2\left(\frac{d}{dy}\right)^2\right\} u_x = V;$$

$$\therefore u = \left\{\left(E_x - \frac{d}{dy}\right)\left(E_x - 2\frac{d}{dy}\right)\right\}^{-1} V$$

$$= \frac{d}{dy}\left(E_x - 2\frac{d}{dy}\right)^{-1} V - \frac{d}{dy}\left(E_x - \frac{d}{dy}\right)^{-1} V$$

$$= \frac{d}{dy}\left(2\frac{d}{dy}\right)^{x-1} \Sigma \left(2\frac{d}{dy}\right)^{-x} V - \frac{d}{dy}\left(\frac{d}{dy}\right)^{x-1} \Sigma \left(\frac{d}{dy}\right)^{-x} V.$$

The complementary part of the value of u introduced by the performance of Σ will evidently be

$$2^x\left(\frac{d}{dy}\right)^x \phi(y) + \left(\frac{d}{dy}\right)^x \psi(y).$$

But in particular cases the difficulties attending the reduction of the general solution may be avoided.

Thus, representing V by V_x, we have, as a particular solution,

$$u_x = \left\{E_x^2 - 3E_x\frac{d}{dy} + 2\left(\frac{d}{dy}\right)^2\right\}^{-1} V_x$$

$$= \left(E_x^{-2} + 3E_x^{-3}\frac{d}{dy} + 7E_x^{-4}\frac{d^2}{dy^2} + \dots\right) V_x$$

$$= V_{x-2} + 3\frac{dV_{x-3}}{dy} + 7\frac{d^2 V_{x-4}}{dy^2} + \dots,$$

which terminates if V_x is rational and integral with respect to y. The complement must then be added.

Thus the complete solution of the given equation when

$$V = F(x) + y,$$

is $$u = F(x-2) + y + 3 + 2^x\left(\frac{d}{dy}\right)^x \phi(y) + \left(\frac{d}{dy}\right)^x \psi(y).$$

EXERCISES.

Solve the equations:

1. $\Delta_x u_{x,y} - a\dfrac{d}{dy}u_{x,y} = 0.$

2. $u_{x+2,y} - a\dfrac{d}{dy}u_{x+1,y} + b\dfrac{d^2}{dy^2}u_{x,y} = 0.$

3. $u_{x+1,y} - u_{x,y+1} = x + y.$

4. $u_{x+1,y+n} - u_{x,y} = a^{x-y}.$

5. $u_{x+2,y} - a^2 u_{x,y+2} = 0.$

6. $u_{x+3,y} - 3u_{x+2,y+1} + 3u_{x+1,y+2} - u_{x,y+3} = x.$

7. $u_{x+1,y+1} - au_{x+1,y} - bu_{x,y+1} + abu_x = c^{x+y}.$

8. $u_{x+3,y} - 3a^2 u_{x+1,y+2} + a^3 u_{x,y+3} = xy.$

9. $\Delta\dfrac{du_x}{dx} - \dfrac{du_x}{dx} - \Delta u_x + u_x = \epsilon^{nx}.$

10. Determine $u_{x,t}$ from the equation

$$c^2\frac{d^2}{dt^2}u_{x+2,t} = \Delta^2 u_{x,t},$$

where Δ affects x only; and, assuming as initial conditions

$$u_{x,0} = ax + b,\quad \frac{d}{d0}u_{x,0} = a'r^x,$$

shew that

$$\frac{d}{dt}u_{x,t} = A\lambda^x(\mu^t + \mu^{-t}),$$

where A, λ and μ are constants (*Cambridge Problems*).

11. Given

$$u_{x+1, y+1} + (a - x - 2y - 2) u_{x, y+1} + (x + y) u_{x,y} = 0$$

with the conditions

$$u_{x,-1} = 0, \; u_{0,0} = 0, \text{ and } u_{x, x+1} = 0,$$

find $u_{x,y}$.

[Cayley, *Tortolini*, Series II. Vol. II. p. 219.]

12. $u_{x,y} = u_{x-y,1} + u_{x-y,2} + \cdots \quad + u_{x-y,y}$.

[De Morgan, *Camb. Math. Jour.* Vol. IV. p. 87.]

CHAPTER XV.

OF THE CALCULUS OF FUNCTIONS.

1. THE calculus of functions in its purest form is distinguished by this, viz. that it recognizes no other operations than those termed functional. In the state to which it has been brought more especially by the labours of Mr Babbage, it is much too extensive a branch of analysis to permit of our attempting here to give more than a general view of its objects and its methods. But it is proper that it should be noticed, 1st, because the Calculus of Finite Differences is but a particular form of the Calculus of Functions; 2ndly, because the methods of the more general Calculus are in part an application, in part an extension of those of the particular one.

In the notation of the Calculus of Functions, $\phi\{\psi(x)\}$ is usually expressed in the form $\phi\psi x$, brackets being omitted except when their use is indispensable. The expressions $\phi\phi x$, $\phi\phi\phi x$ are, by the adoption of indices, abbreviated into $\phi^2 x$, $\phi^3 x$, ... As a consequence of this notation we have $\phi^0 x = x$ independently of the form of ϕ. The inverse form ϕ^{-1} is, it must be remembered, *defined* by the equation

$$\phi\phi^{-1}x = x. \qquad (1).$$

Hence ϕ^{-1} may have different forms corresponding to the same form of ϕ. Thus if

$$\phi x = x^2 + ax,$$

we have, putting $\phi x = t$,

$$x = \phi^{-1}t = -\frac{a \pm \sqrt{(a^2 + 4t)}}{2},$$

and ϕ^{-1} has two forms.

The problems of the Calculus of Functions are of two kinds, viz.

1st. Those in which it is required to determine a functional form equivalent to some known combination of known forms; e.g. from the form of ψx to determine that of $\psi^n x$. This is exemplified in B, page 167.

2ndly. Those which involve the solution of functional equations, i.e. the determination of an unknown function from the conditions to which it is subject, not as in the previous case from the known mode of its composition.

We may properly distinguish these problems as direct and inverse. Problems will of course present themselves in which the two characters meet.

Direct Problems.

2. Given the form of ψx, required that of $\psi^n x$.

There are cases in which this problem can be solved by successive substitution.

Ex. 1. Thus, if $\psi x = x^a$, we have

$$\psi\psi x = (x^a)^a = x^{a^2},$$

and generally

$$\psi^n x = x^{a^n}.$$

Again, if on determining $\psi^2 x$, $\psi^3 x$ as far as convenient it should appear that some one of these assumes the particular form x, all succeeding forms will be determined.

Ex. 2. Thus if $\psi x = 1 - x$, we have

$$\psi^2 x = 1 - (1 - x) = x.$$

Hence $\psi^n x = 1 - x$ or x according as n is odd or even.

Ex. 3. If $\psi x = \dfrac{1}{1-x}$, we find

$$\psi^2 x = \frac{x-1}{x}, \quad \psi^3 x = x.$$

Hence $\psi^n x = x$, $\dfrac{1}{1-x}$ or $\dfrac{x-1}{x}$ according as on dividing n by 3 the remainder is 0, 1 or 2.

Functions of the above class are called periodic, and are distinguished in order according to the number of distinct forms to which $\psi^n x$ gives rise for integer values of n. The function in Ex. 2 is of the second, that in Ex. 3 of the third, order.

Theoretically the solution of the general problem may be made to depend upon that of a difference-equation of the first order by the converse of the process on page 167. For assume

$$\psi^n x = t_n, \quad \psi^{n+1} x = t_{n+1} \qquad (2).$$

Then, since $\psi^{n+1} x = \psi\psi^n x$, we have

$$t_{n+1} = \psi(t_n) \qquad (3).$$

The arbitrary constant in the solution of this equation may be determined by the condition $t_1 = \psi x$, or by the still prior condition

$$t_0 = \psi^0 x = x \qquad (4).$$

It will be more in analogy with the notation of the other chapters of this work if we present the problem in the form: Given ψt, required $\psi^x t$, thus making x the independent variable of the difference-equation.

Ex. 4. Given $\psi t = a + bt$, required $\psi^x t$.

Assuming $\psi^x t = u_x$ we have

$$u_{x+1} = a + bu_x,$$

the solution of which is

$$u_x = cb^x + \frac{a}{1-b}.$$

Now $u_0 = \psi^0 t = t$, therefore

$$t = c + \frac{a}{1-b}.$$

Hence determining c we find on substitution

$$u_x = a\,\frac{b^x - 1}{b - 1} + b^x t \qquad (5),$$

the expression for $\psi^x t$ required.

Ex. 5. Given $\psi t = \dfrac{a}{b+t}$, required $\psi^x t$.

Assuming $\psi^x t = u_x$ we have

$$u_{x+1} = \frac{a}{b + u_x},$$

or

$$u_x u_{x+1} + b u_{x+1} = a.$$

Assuming as in Ch. XII. Art. 1,

$$u_x + b = \frac{v_{x+1}}{v_x},$$

we get

$$v_{x+2} - b v_{x+1} - a v_x = 0,$$

the solution of which is

$$v_x = c_1 \alpha^x + c_2 \beta^x,$$

α and β being the roots of the equation

$$m^2 - bm - a = 0.$$

Hence

$$u_x = \frac{c_1 \alpha^{x+1} + c_2 \beta^{x+1}}{c_1 \alpha^x + c_2 \beta^x} - b;$$

or, putting C for $\dfrac{c_2}{c_1}$ and $\alpha + \beta$ for b, and reducing,

$$u_x = -\alpha\beta\,\frac{\alpha^{x-1} + C\beta^{x-1}}{\alpha^x + C\beta^x} \qquad (6).$$

Now $u_0 = \psi^0 t = t$, therefore

$$t = -\alpha\beta\,\frac{\alpha^{-1} + C\beta^{-1}}{1 + C}$$

$$= -\frac{\beta + C\alpha}{1 + C},$$

whence
$$C = -\frac{t+\beta}{t+\alpha};$$

and, substituting in (6),
$$u_x = -\alpha\beta \frac{\alpha^x - \beta^x + (\alpha^{x-1} - \beta^{x-1})\,t}{\alpha^{x+1} - \beta^{x+1} + (\alpha^x - \beta^x)\,t} \qquad (7),$$

the expression for $\psi^x t$ required.

Since in the above example $\psi t = \frac{a}{b+t}$, we have, by direct substitution,
$$\psi^2 t = \frac{a}{b+\psi t} = \frac{a}{b + \cfrac{a}{b+t}},$$

and continuing the process and expressing the result in the usual notation of continued fractions,
$$\psi^x t = \frac{a}{b+}\,\frac{a}{b+}\,\frac{a}{b+\ldots}\,\frac{a}{b+t},$$

the number of simple fractions being x. Of the value of this continued fraction the right-hand member of (7) is therefore the finite expression. And the method employed shews how the calculus of finite differences may be applied to the finite evaluation of various other functions involving definite repetitions of given functional operations.

Ex. 6*. Given $\psi t = \frac{a+bt}{c+et}$, required $\psi^x t$.

Assuming as before $\psi^x t = u_x$, we obtain as the difference-equation
$$eu_x u_{x+1} + cu_{x+1} - bu_x - a = 0 \qquad (8),$$

and applying to this the same method as before, we find
$$u_x = \frac{\alpha^{x+1} + C\beta^{x+1}}{\alpha^x + C\beta^x} - \frac{c}{e} \qquad (9),$$

α and β being the roots of
$$e^2m^2 - (b+c)\,em + bc - ae = 0 \qquad (10);$$

* See also Hoppe, *Zeitschrift*, v. 136.

and in order to satisfy the condition $u_0 = t$,

$$C = -\frac{e(t-\alpha)+c}{e(t-\beta)+c} \qquad (11).$$

When α and β are imaginary, the exponential forms must be replaced by trigonometrical ones. We may, however, so integrate the equation (8) as to arrive directly at the trigonometrical solution.

For let that equation be placed in the form

$$\left(u_x + \frac{c}{e}\right)\left(u_{x+1} - \frac{b}{e}\right) + \frac{bc-ae}{e^2} = 0.$$

Then assuming $u_x = t_x + \dfrac{b-c}{2e}$, we have

$$\left(t_x + \frac{b+c}{2e}\right)\left(t_{x+1} - \frac{b+c}{2e}\right) + \frac{bc-ae}{e^2} = 0,$$

or

$$t_x t_{x+1} + \mu(t_{x+1} - t_x) + \nu^2 = 0 \qquad (12),$$

in which

$$\mu = \frac{b+c}{2e}, \quad \nu^2 = \frac{bc-ae}{e^2} - \frac{(b+c)^2}{4e^2} \qquad (13).$$

Hence

$$\frac{t_{x+1} - t_x}{\nu^2 + t_x t_{x+1}} = -\frac{1}{\mu},$$

or, assuming $t_x = \nu s_x$,

$$\frac{s_{x+1} - s_x}{1 + s_x s_{x+1}} = \frac{-\nu}{\mu},$$

the integral of which is

$$s_x = \tan\left(C - x\tan^{-1}\frac{\nu}{\mu}\right).$$

But $t_x = \nu s_x$ and $u_x = t_x + \mu'$, where

$$\mu' = \frac{b-c}{2e} \qquad (14).$$

Hence

$$u_x = \nu\tan\left(C - x\tan^{-1}\frac{\nu}{\mu}\right) + \mu' \qquad (15),$$

the general integral.

Now the condition $u_0 = t$ gives

$$t = \nu \tan C + \mu'.$$

Hence determining C we have, finally,

$$\psi^x t = \nu \tan \left(\tan^{-1} \frac{t-\mu'}{\nu} - x \tan^{-1} \frac{\nu}{\mu}\right) + \mu' \qquad (16),$$

for the general expression of $\psi^x t$.

This expression is evidently reducible to the form

$$\frac{A + Bt}{C + Et},$$

the coefficients A, B, C, E being functions of x.

Reverting to the exponential form of $\psi^x t$ given in (9), it appears from (10) that it is real if the function

$$\frac{(b+c)^2}{e^2} - 4\frac{bc - ae}{e^2}$$

is positive. But this is the same as $-4\nu^2$. The trigonometrical solution therefore applies when the expression represented by ν^2 is positive, the exponential one when it is negative.

In the case of $\nu = 0$ the difference-equation (12) becomes

$$t_x t_{x+1} + \mu (t_{x+1} - t_x) = 0,$$

$$\text{or } \frac{1}{t_{x+1}} - \frac{1}{t_x} = \frac{1}{\mu},$$

the integral of which is

$$t_x = \frac{x}{\mu} + C.$$

Determining the constant as before we ultimately get

$$\psi^x t = \frac{\mu'^2 x - (\mu + \mu' x) t}{\mu' x - \mu - xt} \qquad (17),$$

a result which may also be deduced from the trigonometrical solution by the method proper to indeterminate functions.

Periodical Functions.

3. It is thus seen, and it is indeed evident *a priori*, that in the above cases the form of $\psi^x t$ is similar to that of ψt, but with altered constants. The only functions which are known to possess this property are

$$\frac{a+bt}{c+et} \text{ and } at^c.$$

On this account they are of great importance in connexion with the general problem of the determination of the possible forms of periodical functions, particular examples of which will now be given.

Ex. 7. Under what conditions is $a + bt$ a periodical function of the xth order?

By Ex. 4 we have

$$\psi^x t = a\frac{b^x - 1}{b-1} + b^x t,$$

and this, for the particular value of x in question, must reduce to t. Hence

$$a\,\frac{b^x - 1}{b-1} = 0, \quad b^x = 1,$$

equations which require that b should be any x^{th} root of unity except 1 when a is not equal to 0, and any x^{th} root of unity when a is equal to 0.

Hence if we confine ourselves to real forms the only periodic forms of $a + bt$ are t and $a - t$, the former being of every order, the latter of every even order.

Ex. 8. Required the conditions under which $\dfrac{a+bt}{c+et}$ is a periodical function of the xth order.

In the following investigation we exclude the supposition of $e = 0$, which merely leads to the case last considered.

Making then in (16) $\psi^x t = t$, we have

$$t = \mu' + \nu \tan\left(\tan^{-1}\frac{t-\mu'}{\nu} - x\tan^{-1}\frac{\nu}{\mu}\right) \qquad (18),$$

or
$$\frac{t-\mu'}{\nu} = \tan\left(\tan^{-1}\frac{t-\mu'}{\nu} - x\tan^{-1}\frac{\nu}{\mu}\right),$$

an equation which, with the exception of a particular case to be noted presently, is satisfied by the assumption

$$x\tan^{-1}\frac{\nu}{\mu} = i\pi,$$

i being an integer. Hence we have

$$\frac{\nu}{\mu} = \tan\frac{i\pi}{x} \qquad (19),$$

or, substituting for ν and μ their values from (13),

$$\frac{4(bc-ae)}{(b+c)^2} - 1 = \tan^2\frac{i\pi}{x},$$

whence we find

$$e = -\frac{b^2 - 2bc\cos\dfrac{2i\pi}{x} + c^2}{4a\cos^2\dfrac{i\pi}{x}} \qquad (20).$$

The case of exception above referred to is that in which $\nu = 0$, and in which therefore, as is seen from (19), i is a multiple of x. For the assumption $\nu = 0$ makes the expression for t given in (18) indeterminate, the last term assuming the form $0 \times \infty$. If the true limiting value of that term be found in the usual way, we shall find for t the same expression as was obtained in (17) by direct integration. But that expression would lead merely to $x = 0$ as the condition of periodicity, a condition which however is satisfied by all functions whatever, in virtue of the equation $\phi^0 t = t$.

The solution (9) expressed in exponential forms does not lead to any condition of periodicity when a, b, c, e are real quantities.

We conclude that the conditions under which $\dfrac{a+bt}{c+et}$, *when not of the form* $A+Bt$, *is a periodical function of the* x^{th} *order, are expressed by* (20), i *being any integer which is not a multiple of* x*.

4. From any given periodical function an infinite number of others may be deduced by means of the following theorem.

THEOREM. If ft be a periodical function, then $\phi f\phi^{-1}t$ is also a periodical function of the same order

For let $$\phi f\phi^{-1}t = \psi t,$$

then $$\psi^2 t = \phi f\phi^{-1}\phi f\phi^{-1}t$$
$$= \phi f^2\phi^{-1}t.$$

And continuing the process of substitution
$$\psi^n t = \phi f^n \phi^{-1}t.$$

Now, if ft be periodic of the nth order, $f^n t = t$, and
$$f^n\phi^{-1}t = \phi^{-1}t.$$

Hence $$\psi^n t = \phi\phi^{-1}t = t.$$

Therefore ψt is periodic of the nth order.

Thus, it being given that $1-t$ is a periodic function of t of the second order, other such functions are required.

Represent $1-t$ by ft.

Then if $\phi t = t^2$,
$$\phi f\phi^{-1}t = (1-\sqrt{t})^2.$$

If $\phi t = \sqrt{t}$,
$$\phi f\phi^{-1}t = (1-t^2)^{\frac{1}{2}}.$$

These are periodic functions of the second order; and the number might be indefinitely multiplied.

The system of functions included in the general form $\phi f\phi^{-1}t$ have been called the *derivatives* of the function ft.

* I am not aware that the limitation upon the integral values of i has been noticed before. (1st Ed.)

Functional Equations.

5. The most general definition of a functional equation is that it expresses a relation arising from the *forms* of functions; a relation therefore which is independent of the particular values of the subject variable. The object of the solution of a functional equation is the discovery of an unknown form from its relation thus expressed with forms which are known.

The nature of functional equations is best seen from an example of the mode of their genesis.

Let $f(x, c)$ be a given function of x and c, which considered as a function of x, may be represented by ϕx, then

$$\phi x = f(x, c),$$

and changing x into any *given* function ψx,

$$\phi\psi x = f(\psi x, c).$$

Eliminating c between these two equations we have a result of the form

$$F(x, \phi x, \phi\psi x) = 0 \qquad (1).$$

This is a functional equation, the object of the solution of which would be the discovery of the form ϕ, those of F and ψ being given.

It is evident that neither the above process nor its result would be affected if c instead of being a constant were a function of x which did not change its form when x was changed into ψx. Thus if we assume as a primitive equation

$$\phi(x) = cx + \frac{1}{c} \qquad (a),$$

and change x into $-x$, we have

$$\phi(-x) = -cx + \frac{1}{c}.$$

Eliminating c we have, on reduction,

$$\{\phi(x)\}^2 - \{\phi(-x)\}^2 = 4x,$$

a functional equation of which (*a*) constitutes the complete primitive. In that primitive we may however interpret c as an arbitrary *even* function of x, the only condition to which it is subject being that it shall not change on changing x into $-x$. Thus we should have as *particular* solutions

$$\phi(x) = x \cos x + \frac{1}{\cos x},$$

$$\phi(x) = x^3 + \frac{1}{x^2},$$

these being obtained by assuming $c = \cos x$ and x^2 respectively.

Difference-equations are a particular species of functional equations, the elementary functional change being that of x into $x+1$. And the most general method of solving functional equations of all species, consists in reducing them to difference-equations. Laplace has given such a method, which we shall exemplify upon the equation

$$F(x, \phi\psi x, \phi\chi x) = 0 \qquad (2),$$

the forms of ψ and χ being known and that of ϕ sought. But though we shall consider the above equation under its general form, we may remark that it is reducible to the simpler form (1). For, the form of ψ being known, that of ψ^{-1} may be presumed to be known also. Hence if we put $\psi x = z$ and $\chi\psi^{-1}z = \psi_1 z$, we have

$$F(\psi^{-1}z, \phi z, \phi\psi_1 z) = 0,$$

and this, since ψ^{-1} and ψ_1 are known, is reducible to the general form (1).

Now resuming (2) let

$$\left.\begin{array}{ll} \psi x = u_t, & \chi x = u_{t+1} \\ \phi\psi x = v_t, & \phi\chi x = v_{t+1} \end{array}\right\} \qquad (3).$$

Hence v_t and u_t being connected by the relation

$$v_t = \phi u_t \qquad (4),$$

the form of ϕ will be determined if we can express v_t as a function of u_t.

Now the first two equations of the system give on eliminating x a difference-equation of the form

$$u_{t+1} = fu_t \qquad (5),$$

the solution of which will determine u_t, therefore ψx, therefore, by inversion, x as a function of t. This result, together with the last two equations of the system (3), will convert the given equation (2) into a difference-equation of the first order between t and v_t, the solution of which will determine v_t as a function of t, therefore as a function of u_t since the form of u_t has already been determined. But this determination of v_t as a function of u_t is equivalent, as has been seen, to the determination of the form of ϕ.

Ex. 9. Let the given equation be $\phi(mx) - a\phi(x) = 0$.

Then assuming

$$\left.\begin{array}{ll} x = u_t, & mx = u_{t+1} \\ \phi(x) = v_t, & \phi(mx) = v_{t+1} \end{array}\right\} \qquad (a),$$

we have from the first two

$$u_{t+1} - mu_t = 0,$$

the solution of which is

$$u_t = Cm^t \qquad (b).$$

Again, by the last two equations of (a) the given equation becomes

$$v_{t+1} - av_t = 0,$$

whence

$$v_t = C'a^t \qquad (c).$$

Eliminating t between (b) and (c), we have

$$v_t = C'a^{\frac{\log u_t - \log C}{\log m}}.$$

Hence replacing u_t by x, v_t by ϕx, and $C'a^{-\frac{\log C}{\log m}}$ by C_1, we have

$$\phi x = C_1 a^{\frac{\log x}{\log m}} \qquad (d).$$

And here C_1 must be interpreted as any function of x which does not change on changing x into mx.

If we attend strictly to the analytical origin of C_1 in the above solution we should obtain for it the expression

$$a_0 + a_1 \cos\left(2\pi\frac{\log x}{\log m}\right) + a_2 \cos\left(4\pi\frac{\log x}{\log m}\right) + \dots$$

$$+ b_1 \sin\left(2\pi\frac{\log x}{\log m}\right) + b_2 \sin\left(4\pi\frac{\log x}{\log m}\right) + \dots$$

a_0, a_1, b_1, ... being *absolute* constants. But it suffices to adopt the simpler definition given above, and such a course we shall follow in the remaining examples.

Ex. 10. Given $\phi\left(\frac{1+x}{1-x}\right) - a\phi(x) = 0.$

Assuming

$$x = u_t, \quad \frac{1+x}{1-x} = u_{t+1},$$

$$\phi(x) = v_t, \quad \phi\left(\frac{1+x}{1-x}\right) = v_{t+1},$$

we have

$$u_{t+1} = \frac{1+u_t}{1-u_t},$$

or

$$u_t u_{t+1} - u_{t+1} + u_t + 1 = 0.$$

The solution of which is

$$u_t = \tan\left(C + \frac{\pi}{4}t\right).$$

Again we have

$$v_{t+1} - av_t = 0,$$

whence

$$v_t = C'a^t.$$

Hence replacing u_t by x, v_t by $\phi(x)$, and eliminating t,

$$\phi(x) = C_1 a^{\frac{4}{\pi}\tan^{-1}x}.$$

C_1 being any function of x which does not change on changing x into $\frac{1+x}{1-x}$.

6. Linear functional equations of the form

$$\phi\psi^n x + a_1\phi\psi^{n-1}x + a_2\phi\psi^{n-2}x \ldots + a_n\phi(x) = X \qquad (6),$$

where $\psi(x)$ is a known function of x, may be reduced to the preceding form.

For let π be a symbol which operating on any function $\phi(x)$ has the effect of converting it into $\phi\psi(x)$. Then the above equation becomes

$$\pi^n\phi(x) + a_1\pi^{n-1}\phi(x) \ldots + a_n\phi(x) = X,$$

or

$$(\pi^n + a_1\pi^{n-1} \ldots + a_n)\,\phi(x) = X \qquad (7).$$

It is obvious that π possesses the distributive property expressed by the equation

$$\pi(u+v) = \pi u + \pi v,$$

and that it is commutative with constants so that

$$\pi a u = a\pi u.$$

Hence we are permitted to reduce (7) in the following manner, viz.

$$\begin{aligned}\phi(x) &= (\pi^n + a_1\pi^{n-1} \ldots + a_n)^{-1}X \\ &= \{N_1(\pi - m_1)^{-1} + N_2(\pi - m_2)^{-1} \ldots\}\,X \qquad (8),\end{aligned}$$

$m_1, m_2 \ldots$ being the roots of

$$m^n + a_1 m^{n-1} \ldots + a_n = 0 \qquad (9),$$

and $N_1, N_2 \ldots$ having the same values as in the analogous resolution of rational fractions.

Now if $(\pi - m)^{-1}X = \phi(x)$, we have

$$(\pi - m)\,\phi(x) = X,$$

or
$$\phi\psi(x) - m\phi(x) = X,$$
to which Laplace's method may be applied.

Ex. 11. Given $\phi(m^2x) + a\phi(mx) + b\phi(x) = x^n$.

Representing by α and β the roots of $x^2 + ax + b = 0$, the solution is

$$\phi(x) = \frac{x^n}{m^{2n} + am^n + b} + Cx^{\frac{\log\alpha}{\log m}} + C'x^{\frac{\log\beta}{\log m}},$$

C and C' being functions of x unaffected by the change of x into mx.

Here we may notice that just as in linear differential equations and in linear difference-equations, and for the same reason, viz. the distributive character of the symbol π, the complete value of $\phi(x)$ consists of two portions, viz. of any particular value of $\phi(x)$ together with what would be its complete value where $X = 0$. This is seen in the above example.

7. There are some cases in which particular solutions of functional equations, more especially if the known functions involved in the equations are periodical, may be obtained with great ease. The principle of their solution is as follows.

Supposing the given equation to be

$$F(x, \phi x, \phi\psi x) = 0 \qquad (10),$$

and let ψx be a periodical function of the second order. Then changing x into ψx, and observing that $\psi^2 x = x$, we have

$$F(\psi x, \phi\psi x, \phi x) = 0 \qquad (11).$$

Eliminating $\phi\psi x$ the resulting equation will determine ϕx as a function of x and ψx, and therefore since ψx is supposed known, as a function of x.

If ψx is a periodical function of the third order, it would be necessary to effect the substitution twice in succession, and then to eliminate $\phi\psi x$, and $\phi\psi^2 x$; and so on according to the order of periodicity of ψx.

Ex. 12. Given $$(\phi x)^2 \phi \frac{1-x}{1+x} = a^2 x.$$

The function $\frac{1-x}{1+x}$ is periodic of the second order. Change then x into $\frac{1-x}{1+x}$, and we have

$$\left(\phi \frac{1-x}{1+x}\right)^2 \phi x = a^2 \frac{1-x}{1+x}.$$

Hence, eliminating $\phi \frac{1-x}{1+x}$, we find

$$\phi x = a^{\frac{2}{3}} x^{\frac{1}{3}} \left(\frac{1+x}{1-x}\right)^{\frac{1}{3}}$$

as a particular solution. (Babbage, *Examples of Functional Equations*, p. 7.)

This method fails if the process of substitution does not yield a number of *independent* equations sufficient to enable us to effect the elimination. Thus, supposing ψx a periodical function of the second order, it fails for equations of the form

$$F(\phi x, \phi\psi x) = 0,$$

if symmetrical with respect to ϕx and $\phi\psi x$. In such cases we must either, with Mr Babbage, treat the given equation as a particular case of some more general equation which is unsymmetrical, or we must endeavour to solve it by some more general method like that of Laplace.

Ex. 13. Given

$$(\phi x)^2 + \left\{\phi\left(\frac{\pi}{2} - x\right)\right\}^2 = 1.$$

This is a particular case of the more general equation

$$(\phi x)^2 + m\left\{\phi\left(\frac{\pi}{2} - x\right)\right\}^2 = 1 + n\chi x,$$

m and n being constants which must be made equal to 1 and 0 respectively, and χx being an arbitrary function of x.

Changing x into $\frac{\pi}{2} - x$, we have

$$\left\{\phi\left(\frac{\pi}{2} - x\right)\right\}^2 + m\{\phi(x)\}^2 = 1 + n\chi\left(\frac{\pi}{2} - x\right).$$

Eliminating $\phi\left(\frac{\pi}{2} - x\right)$ from the above equations we find

$$(1 - m^2)\{\phi(x)\}^2 = 1 - m + n\left\{\chi x - m\chi\left(\frac{\pi}{2} - x\right)\right\}.$$

Therefore

$$\{\phi(x)\}^2 = \frac{1}{1+m} + \frac{n}{1-m^2}\left\{\chi x - m\chi\left(\frac{\pi}{2} - x\right)\right\}.$$

Now if m become 1 and n become 0, independently, the fraction $\frac{n}{1-m^2}$ becomes indeterminate, and may be replaced by an arbitrary constant c. Thus we have

$$\{\phi(x)\}^2 = \frac{1}{2} + c\chi(x) - c\chi\left(\frac{\pi}{2} - x\right);$$

whence, merging c in the arbitrary function,

$$\phi(x) = \left\{\frac{1}{2} + \chi(x) - \chi\left(\frac{\pi}{2} - x\right)\right\}^{\frac{1}{2}} \qquad (12).$$

The above is in effect Mr Babbage's solution, excepting that, making m and n dependent, he finds a particular value for the fraction which in the above solution becomes an arbitrary constant.

Let us now solve the equation by Laplace's method. Let $\{\phi(x)\}^2 = \psi x$, and we have

$$\psi(x) + \psi\left(\frac{\pi}{2} - x\right) = 1.$$

Hence assuming

$$x = u_t, \quad \frac{\pi}{2} - x = u_{t+1},$$

$$\psi(x) = v_t, \quad \psi\left(\frac{\pi}{2} - x\right) = v_{t+1},$$

we have

$$u_{t+1} + u_t = \frac{\pi}{2},$$

$$v_{t+1} + v_t = 1.$$

The solutions of which are

$$u_t = c_1 (-1)^x + \frac{\pi}{4},$$

$$v_t = c_2 (-1)^x + \frac{1}{2}.$$

Hence

$$\frac{v_t - \frac{1}{2}}{u_t - \frac{\pi}{4}} = \frac{c_2}{c_1} = C.$$

Therefore

$$v_t = \frac{1}{2} + C\left(u_t - \frac{\pi}{4}\right),$$

or

$$\psi(x) = \frac{1}{2} + C\left(x - \frac{\pi}{4}\right).$$

Therefore

$$\phi(x) = \left\{\frac{1}{2} + C\left(x - \frac{\pi}{4}\right)\right\}^{\frac{1}{2}},$$

in which C must be interpreted as a function of x which does not change when x is changed into $\frac{\pi}{2} - x$. It is in fact *an arbitrary symmetrical function of x and $\frac{\pi}{2} - x$.*

The previous solution (12) is included in this.

For, equating the two values of $\phi(x)$ with a view to determine C, we find

$$C = \frac{\chi(x) - \chi\left(\frac{\pi}{2} - x\right)}{x - \frac{\pi}{4}}$$

$$= \frac{\chi(x)}{x - \frac{\pi}{4}} + \frac{\chi\left(\frac{\pi}{2} - x\right)}{\frac{\pi}{2} - x - \frac{\pi}{4}},$$

which is seen to be symmetrical with respect to x and $\frac{\pi}{2} - x$.

8. There are certain equations, and those of no inconsiderable importance, which involve at once two independent variables in such functional connexion that by differentiation and elimination of one or more of the functional terms, the solution will be made ultimately to depend upon that of a differential equation.

Ex. 14. Representing by $P\phi(x)$ the unknown magnitude of the resultant of two forces, each equal to P, acting in one plane and inclined to each other at an angle $2x$, it is shewn by Poisson (*Mécanique*, Tom. I. p. 47) that on certain assumed principles, viz. the principle that the order in which forces are combined into resultants is indifferent—the principle of (so-called) sufficient reason, ..., the following functional equation will exist independently of the particular values of x and y, viz.

$$\phi(x+y) + \phi(x-y) = \phi(x)\,\phi(y).$$

Now, differentiating twice with respect to x, we have

$$\phi''(x+y) + \phi''(x-y) = \phi''(x)\,\phi(y).$$

And differentiating the same equation twice with respect to y,

$$\phi''(x+y) + \phi''(x-y) = \phi(x)\,\phi''(y).$$

Hence
$$\frac{\phi''(x)}{\phi(x)} = \frac{\phi''(y)}{\phi(y)}.$$

Thus the value of $\dfrac{\phi''(x)}{\phi(x)}$ is quite independent of that of x. We may therefore write

$$\frac{\phi''(x)}{\phi(x)} = \pm m^2,$$

m being an arbitrary constant. The solution of this equation is

$$\phi(x) = A\epsilon^{mx} + B\epsilon^{-mx}, \text{ or } \phi(x) = A \cos mx + B \sin mx.$$

Substituting in the given equation to determine A and B, we find

$$\phi(x) = \epsilon^{mx} + \epsilon^{-mx}, \quad \text{or } 2 \cos mx.$$

Now assuming, on the afore-named principle of sufficient reason, that three equal forces, each of which is inclined to the two others at angles of 120°, produce equilibrium, it follows that $\phi\left(\dfrac{\pi}{3}\right) = 1$. This will be found to require that the second form of $\phi(x)$ be taken, and that m be made equal to 1. Thus $\phi(x) = 2 \cos x$. And hence the known law of composition of forces follows.

Ex. 15. A ball is dropped upon a plane with the intention that it shall fall upon a given point, through which two perpendicular axes x and y are drawn. Let $\phi(x)\,dx$ be the probability that the ball will fall at a distance between x and $x + dx$ from the axis y, and $\phi(y)\,dy$ the probability that it will fall at a distance between y and $y + dy$ from the axis x. Assuming that the tendencies to deviate from the respective axes are independent, what must be the form of the function $\phi(x)$ in order that the probability of falling upon any particular point of the plane may be independent of the position of the rectangular axes? (Herschel's *Essays.*)

The functional equation is easily found to be

$$\phi(x)\,\phi(y) = \phi\{\sqrt{(x^2 + y^2)}\}\,\phi(0).$$

Differentiating with respect to x and with respect to y, we have

$$\phi'(x)\,\phi(y) = \frac{x\phi'\{\sqrt{(x^2+y^2)}\}\,\phi(0)}{\sqrt{(x^2+y^2)}},$$

$$\phi(x)\,\phi'(y) = \frac{y\phi'\{\sqrt{(x^2+y^2)}\}\,\phi(0)}{\sqrt{(x^2+y^2)}}.$$

Therefore
$$\frac{\phi'(x)}{x\phi(x)} = \frac{\phi'(y)}{y\phi(y)}.$$

Hence we may write

$$\frac{\phi'(x)}{x\phi(x)} = 2m,$$

a differential equation which gives

$$\phi(x) = C\epsilon^{mx^2}.$$

The condition that $\phi(x)$ must diminish as the absolute value of x increases shews that m must be negative. Thus we have

$$\phi(x) = C\epsilon^{-h^2x^2}.$$

EXERCISES.

1. If $\phi(x) = \dfrac{2x}{1-x^2}$, determine $\phi^n(x)$.

2. If $\phi(x) = 2x^2 - 1$, determine $\phi^n(x)$.

3. If $\psi(t) = \dfrac{a+bt}{c+et}$ and $\psi^x(t) = \dfrac{A+Bt}{C+Et}$, shew, by means of the necessary equation $\psi\psi^x(t) = \psi^x\psi(t)$, that

$$\frac{A}{a} = \frac{E}{e} = \frac{C-B}{c-b}.$$

4. Shew hence that $\psi^x(t)$ may be expressed in the form

$$\frac{a+b_x t}{b_x - b + c + et},$$

the equation for determining b_x being

$$b_x b_{x+1} + c b_{x+1} - b b_x - ae = 0,$$

and that results equivalent to those of Ex. 5, Art. 2, may hence be deduced.

Solve the equations

5. $f(x) + f(y) = f(x+y)$.

6. $f(x) + af(-x) = x^n$.

7. $f(x) - af(-x) = \epsilon^x$.

8. $f(1-x) + f(1+x) = 1 - x^2$.

9. $f(x) = xf'(x) + f\{f'(x)\}$.

10. Find the value, to x terms, of the continued fraction

$$\frac{2}{1+\frac{2}{1+\dots}}$$

11. What particular solution of the equation

$$f(x) + f\left(\frac{1}{x}\right) = a$$

is deducible by the method of Art. 7 from the equation

$$f(x) + mf\left(\frac{1}{x}\right) = a + n\phi(x)\,?$$

12. Required the equation of that class of curves in which the product of any two ordinates, equidistant from a certain ordinate whose abscissa a is given, is equal to the square of that abscissa.

13. If πx be a periodical function of x of the nth degree, shew that there will exist a particular value of $f(\pi)\, x$ expressible in the form

$$a_0 + a_1 \pi x + a_2 \pi^2 x \ldots + a_{n-1} \pi^{n-1} x,$$

and shew how to determine the constants $a_0, a_1, a_2, \ldots, a_{n-1}$.

14. Shew hence that a particular integral of the equation

$$\phi\left(\frac{1+x}{1-x}\right) - a\phi(x) = x$$

will be

$$\phi(x) = \frac{a^3}{1-a^4}\left(x + \frac{1}{a}\frac{1+x}{1-x} - \frac{1}{a^2 x} + \frac{1}{a^3}\frac{x-1}{x+1}\right).$$

15. The complete solution of the above equation will be obtained by adding to the particular value of x the complementary function $Ca^{\frac{4\tan^{-1}x}{\pi}}$

16. Solve the simultaneous functional equations

$$\phi(x+y) = \phi(x) + \frac{\phi(y)\{\psi(x)\}^2}{1-\phi(x)\phi(y)},$$

$$\psi(x+y) = \frac{\psi(x)\psi(y)}{1-\phi(x)\phi(y)}.$$

(*Smith's Prize Examination*, 1860.)

17. Solve the equation

$$nF(nx) = f(x) + f\left(x+\frac{1}{n}\right) + f\left(x+\frac{2}{n}\right) + \cdots + f\left(x+\frac{n-1}{n}\right).$$

[Kinkelin, *Grunert*, XXII. 189.]

18. Solve the equation

$$\phi(x) + \phi(y) = \psi\{xf(y) + yf(x)\}.$$

[Abel, *Crelle*, II. 386.]

Magnus (*Crelle*, v. 365) and Lottner (*Crelle*, XLVI.) have continued the investigations into this and kindred functional equations.

19. Find the conditions that $\phi(x, y) + \sqrt{-1}\,\psi(x, y)$ may be of the form $F(x + y\sqrt{-1})$.

[Dienger, *Grunert*, X. 422.]

20. Shew that

$$z_n = \frac{d^n u}{dx^n} \div \frac{d^{n-1}u}{dx^{n-1}}$$

satisfies the equation

$$\frac{dz_n}{dx} = z_n \Delta z_n,$$

u being any function of x.

If a regular polygon, which is inscribed in a fixed circle, be moveable, and if x denote the variable arc between one of its angles and a fixed point in the circumference, and z_n the ratio, multiplied by a certain constant, of the distances from the centre of the feet of perpendiculars drawn from the nth and $(n-1)^{\text{th}}$ angles, counting from A, on the diameter through the fixed point, prove that z_n is a function which satisfies the equation.

21. If $\phi(z) = \phi(x)\,\phi(y)$, where z is a function of x and y determined by the equation $f(z) = f(x) f(y)$, find the form of $\phi(x)$.

CHAPTER XVI.

GEOMETRICAL APPLICATIONS.

1. THE determination of a curve from some property connecting points separated by finite intervals usually involves the solution of a difference-equation, pure or mixed, or more generally of a functional equation.

The particular species of this equation will depend upon the law of succession of the points under consideration, and upon the nature of the elements involved in the expression of the given connecting property.

Thus if the abscissæ of the given points increase by a constant difference, and if the connecting property consist merely in some relation between the successive ordinates, the determination of the curve will depend on the integration of a pure difference-equation. But if, the abscissæ still increasing by a constant difference, the connecting property consist in a relation involving such elements as the tangent, the normal, the radius of curvature, ..., the determining equation will be one of mixed differences.

If, instead of the abscissa, some other element of the curve is supposed to increase by a constant difference, it is necessary to assume that element as the independent variable. But when no obvious element of the curve increases by a constant difference, it becomes necessary to assume as independent variable the index of that operation by which we pass from point to point of the curve, i.e. some number which is supposed to measure the frequency of the operation, and which increases by unity as we pass from any point to the succeeding point. Then we must endeavour to form two difference-equations, pure or mixed, one from the law of succession of the points, the other from their connecting property; and from the integrals eliminate the new variable.

There are problems in the expression of which we are led to what may be termed functional differential equations, i.e. equations in which the operation of differentiation and an unknown functional operation seem inseparably involved. In some such cases a procedure similar to that employed in the solution of Clairaut's differential equation enables us to effect the solution.

2. The subject can scarcely be said to be an important one, and a single example in illustration of each of the different kinds of problems, as classified above, may suffice.

Ex. 1. To find a curve such that, if a system of n right lines, originating in a fixed point and terminating in the curve, revolve about that point making always equal angles with each other, their sum shall be invariable. (Herschel's *Examples*, p. 115.)

The angles made by these lines with some fixed line may be represented by

$$\theta,\ \theta+\frac{2\pi}{n},\ \theta+\frac{4\pi}{n},\ldots,\theta+\frac{2(n-1)\pi}{n}.$$

Hence, if $r=\phi(\theta)$ be the polar equation of the curve, the given point being pole, we have

$$\phi(\theta)+\phi\left(\theta+\frac{2\pi}{n}\right)\ldots+\phi\left\{\theta+\frac{2(n-1)\pi}{n}\right\}=na,$$

a being some given quantity.

Let $\theta=\dfrac{2\pi z}{n}$, and let $\phi\left(\dfrac{2\pi z}{n}\right)=u_z$, then we have

$$u_z+u_{z+1}\ldots+u_{z+n-1}=na,$$

the complete integral of which is

$$u_z=a+C_1\cos\frac{2\pi z}{n}+C_2\cos\frac{4\pi z}{n}\ldots+C_{n-1}\cos\frac{(2n-2)\pi z}{n}.$$

Hence we find

$$r = a + C_1 \cos\theta + C_2 \cos 2\theta \ldots + C_{n-1} \cos(n-1)\theta,$$

the analytical form of any coefficient C_i being

$$C_i = A + B_1 \cos n\theta + B_2 \cos 2n\theta + \ldots,$$
$$+ E_1 \sin n\theta + E_2 \sin 2n\theta + \ldots,$$

$A, B_1, E_1, \ldots,$ being absolute constants.

The particular solution $r = a + b\cos\theta$ gives, on passing to rectangular co-ordinates,

$$(x^2 - bx + y^2)^2 = a^2(x^2 + y^2),$$

and the curve is seen to possess the property that "if a system of any number of radii terminating in the curve and making equal angles with each other be made to revclve round the origin of co-ordinates their sum will be invariable."

Ex. 2. Required the curve in which, the abscissæ increasing by a constant value unity, the subnormals increase in a constant ratio 1 : a.

Representing by y_x the ordinate corresponding to the abscissa x, we shall have the mixed difference-equation

$$y_x \frac{dy_x}{dx} - ay_{x-1}\frac{dy_{x-1}}{dx} = 0 \qquad (1).$$

Let $y_x \dfrac{dy_x}{dx} = u_x$, then

$$u_x - au_{x-1} = 0;$$
$$\therefore\ u_x = Ca^x,$$

whence

$$y_x \frac{dy_x}{dx} = Ca^x \qquad (2).$$

Hence integrating we find

$$y_x = \sqrt{(C_1 a^x + c)} \qquad (3),$$

C_1 being a periodical constant which does not vary when x changes to $x+1$, and c an absolute constant.

Ex. 3. Required a curve such that a ray of light proceeding from a given point in its plane shall after two reflections by the curve return to the given point.

The above problem has been discussed by Biot, whose solution as given by Lacroix (*Diff. and Int. Calc.* Tom. III. p. 588) is substantially as follows:

Assume the given radiant point as origin; let x, y be the co-ordinates of the first point of incidence on the curve, and x', y' those of the second. Also let $\frac{dy}{dx}=p$, $\frac{dy'}{dx'}=p'$.

It is easily shewn that twice the angle which the normal at any point of the curve makes with the axis of x is equal to the sum of the angles which the incident and the corresponding reflected ray at that point make with the same axis.

Now the tangent of the angle which the incident ray at the point x, y makes with the axis of x is $\frac{y}{x}$. The tangent of the angle which the normal makes with the axis of x is $-\frac{1}{p}$, and the tangent of twice that angle is

$$\frac{-\frac{2}{p}}{1-\frac{1}{p^2}}=\frac{2p}{1-p^2}.$$

Hence the tangent of the angle which the ray reflected from x, y makes with the axis of x is

$$\frac{\frac{2p}{1-p^2}-\frac{y}{x}}{1+\frac{2p}{1-p^2}\frac{y}{x}}=\frac{2xp-y(1-p^2)}{x(1-p^2)+2py} \qquad (1).$$

Again, by the conditions of the problem a ray incident from the origin upon the point x', y' would be reflected in the *same* straight line, only in an opposite direction. But the two expressions for the tangent of inclination of the reflected ray being equal,

$$\frac{2x'p' - y'(1-p'^2)}{x'(1-p'^2) + 2y'p'} - \frac{2xp - y(1-p^2)}{x(1-p^2) + 2yp} = 0 \qquad (2),$$

while for the equation of that ray, we have

$$y' - y = \frac{2xp - y(1-p^2)}{x(1-p^2) + 2yp}(x' - x) \qquad (3).$$

Now, regarding x and y as functions of an independent variable z which changes to $z+1$ in passing from the first point of incidence to the second, the above equations become

$$\Delta \frac{2xp - y(1-p^2)}{x(1-p^2) + 2yp} = 0,$$

$$\Delta y = \frac{2xp - y(1-p^2)}{x(1-p^2) + 2yp} \Delta x.$$

The first of these equations gives

$$\frac{2xp - y(1-p^2)}{x(1-p^2) + 2yp} = C \qquad (4),$$

whence by substitution

$$\Delta y = C\Delta x.$$

Therefore

$$y = Cx + C'.$$

Here C and C' are primarily periodic functions of z which do not change when z becomes $z+1$. Biot observes that, if C be such a function, $\phi(C)$, in which the form of ϕ is arbitrary, will also be such, and that we may therefore assume $C' = \phi(C)$, whence

$$y = Cx + \phi(C),$$

and, restoring to C its value in terms of x, y, and p given in (4), we shall have

$$y = x\frac{2xp - y(1-p^2)}{x(1-p^2) + 2yp} + \phi\left\{\frac{2xp - y(1-p^2)}{x(1-p^2) + 2yp}\right\}^* \qquad (5).$$

This is the differential equation of the curve.

Although Lacroix does not point out any restriction on the form of the function ϕ, it is clear that it cannot be quite arbitrary. For if $C = \psi(z)$, we should have

$$C' = \phi\psi(z),$$

and then, giving to ϕ some functional form to which ψ is inverse, there would result

$$C' = z,$$

so that C' would change when z was changed into $z+1$. From the general form of periodic constants, Chap. IV., it is evident that a rational function of such a constant possesses the same character. Thus the differential equation (5) is applicable when ϕ indicates a rational function, and generally when it denotes a functional operation which while periodical itself does not affect the periodical character of its subject.

If we make the arbitrary function 0, we have on reduction

$$(y^2 - x^2)p + xy(1-p^2) = 0,$$

the integral of which is

$$x^2 + y^2 = r^2,$$

denoting a circle.

* It is only while writing this Chapter that a general interpretation of this equation has occurred to me. Its complete primitive denotes a family of curves defined by the following property, viz. that the caustic into which each of these curves would reflect rays issuing from the origin would be identical with the envelope of the system of straight lines defined by the equation $y = cx + \phi(c)$, c being a variable parameter. This interpretation, which is quite irrespective of the form of the function ϕ, confirms the observation in the text as to the necessity of restricting the form of that function in the problem there discussed. I regret that I have not leisure to pursue the inquiry.

I have also ascertained that the differential equation always admits of the following particular solution, viz.

$$(y-A)^2 + (x-B)^2 = 0,$$

A and B being given by the equation

$$\phi(\sqrt{-1}) = A - B\sqrt{-1}. \quad \textit{(1st edition.)}$$

If we make the arbitrary function a constant and equal to $2a$, we find on reduction

$$\{x^2 - (y-a)^2 + a^2\} p - x(y-a)(1-p^2) = 0,$$

the complete primitive of which (*Diff. Equations*, p. 135) is

$$(y-a)^2 + c^2x^2 = \frac{a^2c^2}{1-c^2},$$

the equation of an ellipse about the focus.

3. The following once famous problem engaged in succession the attention of Euler, Biot, and Poisson. But the subjoined solution, which alone is characterized by unity and completeness, is due to the late Mr Ellis, *Cambridge Journal*, Vol. III. p. 131. It will be seen that the problem leads to a functional differential equation.

Ex. 4. Determine the class of curves in which the square of any normal exceeds the square of the ordinate erected at its foot by a constant quantity a.

If $y^2 = \psi(x)$ be the equation of the curve, the subnormal will be $\frac{\psi'(x)}{2}$, and the normal squared $\psi(x) + \left\{\frac{\psi'(x)}{2}\right\}^2$. The equation of the problem will therefore be

$$\psi(x) + \left\{\frac{\psi'(x)}{2}\right\}^2 - \psi\left\{x + \frac{\psi'(x)}{2}\right\} = a \qquad (1).$$

Differentiating, we have

$$\psi'(x) + \psi(x)\frac{\psi''(x)}{2} + \psi'\left\{x + \frac{\psi'(x)}{2}\right\}\left(1 + \frac{\psi''(x)}{2}\right) = 0,$$

which is resolvable into the two equations,

$$1 + \frac{\psi''(x)}{2} = 0 \qquad (2),$$

$$\psi'(x) + \psi'\left\{x + \frac{\psi'(x)}{2}\right\} = 0 \qquad (3).$$

The first of these gives on integration

$$\psi(x) + x^2 = ax + \beta \qquad (4).$$

Substituting the value of $\psi(x)$, hence deduced, in (1), we find as an equation of condition

$$a = 0,$$

and, supposing this satisfied, (4) gives

$$y^2 + x^2 = ax + \beta,$$

the equation of a circle whose centre is on the axis of x. It is evident that this is a solution of the problem, supposing

$$a = 0.$$

To solve the second equation (3), assume

$$x + \tfrac{1}{2}\psi'(x) = \chi(x),$$

and there results

$$\chi^2(x) - 2\chi(x) + x = 0 \qquad (5).$$

To integrate this let $x = u_t$, $\chi(x) = u_{t-1}$, and we have

$$u_{t+2} - 2u_{t+1} + u_t = 0,$$

whence

$$u_t = C + C't,$$

C and C' being functions which do not change on changing t into $t+1$. If we represent them by $P(t)$ and $P_1(t)$, we have

$$u_t = P(t) + tP_1(t),$$

$$u_{t+1} = P(t) + (t+1)P_1(t),$$

whence, since $u_t = x$ and $u_{t+1} = \chi(x) = x + \frac{1}{2}\psi'(x)$,

we have

$$x = P(t) + tP_1(t),$$

$$\tfrac{1}{2}\psi'(x) = P_1(t).$$

Hence

$$\psi'(x)\,dx = P_1(t)\,\{P'(t) + P_1(t) + tP_1'(t)\}\,dt,$$

$$\psi(x) = \int P_1 t\,\{P'(t) + P_1(t) + tP_1'(t)\}\,dt.$$

Replacing therefore $\psi(x)$ by y^2, the solution is expressed by the two equations,

$$\left.\begin{aligned} x &= P(t) + tP_1(t) \\ y^2 &= \int P_1(t)\,\{P'(t) + P_1(t) + tP_1'(t)\}\,dt \end{aligned}\right\} \qquad (6),$$

from which, when the forms of $P(t)$ and $P_1(t)$ are assigned, t must be eliminated.

If we make $P(t) = \alpha$, $P_1(t) = \beta$, thus making them constant, we have

$$x = \alpha + \beta t,$$

$$y^2 = \int \beta^2 dt = \beta^2 t + c.$$

Therefore eliminating t and substituting e for $c - \alpha\beta$,

$$y^2 = \beta x + e.$$

Substituting this in (1), we find

$$\frac{-\beta^2}{4} = a.$$

Thus, in order that the solution should be real, a must be negative. Let $a = -h^2$, then $\beta = \pm 2h$, and

$$y^2 = \pm 2hx + e \qquad (7),$$

the solution required. This indicates two parabolas.

If $a = 0$, the solution represents two straight lines parallel to the axis of x.

EXERCISES.

1. Find the general equation of curves in which the diameter through the origin is constant in value.

2. Find the general equation of the curve in which the product of two segments of a straight line drawn through a fixed point in its plane to meet the curve shall be invariable.

3. If in Ex. 4 of the above Chapter the radiant point be supposed infinitely distant, shew that the equation of the reflecting curve will be of the form

$$y = \frac{2px}{1-p^2} + \phi\left(\frac{2p}{1-p^2}\right),$$

ϕ being restricted as in the Example referred to.

4. If a curve be such that a straight line cutting it perpendicularly at one point shall also cut it perpendicularly at another, prove that the differential equation of the curve will be

$$y = \frac{-x}{p} + \phi\left(\frac{-1}{p}\right),$$

ϕ being restricted as in Ex. 4 of this Chapter.

5. Shew that the integral of the above differential equation, when the form of ϕ is unrestricted, may be interpreted by the system of involutes to the curve which is the envelope of the system of straight lines defined by the equation

$$y = mx + \phi(m),$$

m being a variable parameter.

ANSWERS TO THE EXAMPLES.

CHAPTER II.

6. Obtained from the identity $\Delta^n (0-1)(0-2) \ldots\ldots (0-n) 0^x = 0$.

9. $\epsilon \left(1 + t + t^2 + \frac{5}{6} t^3\right)$.

14. $(x - 6x^3) \cos x - (7x^2 - x^4) \sin x$.

16. (2) $u_x = \dfrac{f(E)}{\lfloor x} 0^x$.

CHAPTER III.

1. 2·3263359, which is correct to the last figure.

2. $x^3 - 9x^2 + 17x + 6$.

3. $v_2 = \dfrac{-3v_0 + 10v_1 + 5v_4 - 2v_5}{10}$, $v_3 = \dfrac{-2v_0 + 5v_1 + 10v_4 - 3v_5}{10}$.

13. It will be so if $\phi(x) = 0$ have one root, and $\phi'(x) = 0$ have no root between 1 and k.

CHAPTER IV.

1. (1) $\dfrac{(2n-1)(2n+1)(2n+3)(2n+5)(2n+7)}{10} + \dfrac{21}{2}$.

(2) $\dfrac{1}{90} - \dfrac{1}{6(2n+1)(2n+3)(2n+5)}$.

(3) $\dfrac{(2n-1)(2n+1)(2n+3)(2n+5)(8n+43)}{40} + \dfrac{129}{8}$.

(4) $\frac{11}{12} - \frac{4n+11}{4(2n+1)(2n+3)}$.

(5) Apply the method of Ex. 8.

(6) Write $2\cos\theta = x + \frac{1}{x}$ and use (10) page 73.

3. $n(2n+1)(8n^2+4n-7)$.

4. $$\frac{\sin\frac{m}{2}(2x-1)}{2\sin\frac{m}{2}}\phi(x)+\frac{\cos\frac{m}{2}(2x)}{\left(2\sin\frac{m}{2}\right)^2}\Delta\phi(x)-\frac{\sin\frac{m}{2}(2x+1)}{\left(2\sin\frac{m}{2}\right)^3}\Delta^2\phi(x)$$

$$-\frac{\cos\frac{m}{2}(2x+2)}{\left(2\sin\frac{m}{2}\right)^4}\Delta^3\phi(x)+\frac{\sin\frac{m}{2}(2x+3)}{\left(2\sin\frac{m}{2}\right)^5}\Delta^4\phi(x)+\dots.$$

6. (1) $\cot\frac{\theta}{2} - \cot 2^{n-1}\theta$.

(2) $\frac{2\sin n\theta}{\cos(n+1)\theta\sin 2\theta}$.

7. $\tan^{-1}(n-1)x + C$, $C - \frac{\log 2\sin 2^n\theta}{2^{n-1}}$, $C + \frac{2^n}{n}$.

8. Assume for the form of the integral

$$\frac{(A+Bx+\dots+Mx^{n-1})s^x}{u_x u_{x+1}\dots u_{x+m-2}},$$

and then seek to determine the constants.

CHAPTER V.

1. $$C - \frac{1}{16}\left\{\frac{1}{\left(n+\frac{1}{4}\right)} + \frac{1}{2\left(n+\frac{1}{4}\right)} + \frac{1}{6\left(n+\frac{1}{4}\right)^3} - \frac{1}{30\left(n+\frac{1}{4}\right)^5}\right\},$$

where $C = 1{\cdot}0787$ approximately and is the sum *ad inf.*

2. $\Sigma \frac{1}{x^5} = C - \frac{1}{4x^4} - \frac{1}{2x^5} - \frac{5}{12x^6} + \frac{7}{24x^8} - \dots$.

The sum *ad inf.* differs from that of the first nine terms by ·0000304167.

3. $\pi^{-\frac{1}{2}} \left\{ 2x^{\frac{1}{2}} + \frac{3}{4x^{\frac{1}{2}}} \right\}$.

4. See page 71.

5. (1) Apply Prop. IV. page 99. If $-a^2$ be written for x^2 in the first series it can be divided into two series similar to the Example there given.

(2) $$\frac{1}{2} \left\{ \frac{1}{x(x+1)(x+2)} + \frac{1}{2} \frac{3}{x(x+1)(x+2)(x+3)} + \frac{1}{4} \frac{3\,4}{x(x+1)(x+2)(x+3)(x+4)} + \dots \right\}.$$

8. $\frac{(z+1)^{(n)} - x^{(n)}}{(z-x+1)\, z^{(n-1)}}$.

13. See Ex. 7. Also page 115.

CHAPTER VII.

1. $\frac{1}{a} \tan^{-1} a$ and $\frac{\pi}{2a}$.

3. (1) Divergent. (2) Convergent.

(3) The successive tests corresponding to (C) are obtained by writing $-\Delta u_{x+n}$ for $\frac{u_x}{u_{x+1}} - 1$ therein. The set corresponding to (B) are obtained by writing

$$-Du_{x+n} \text{ for } \frac{\phi'(x)}{\phi(x)}.$$

(4) Convergent if x be positive, divergent if it be negative.

(5) Divergent. (6) Divergent.

(7) Divergent unless a be greater than unity.

(8) Divergent unless a be greater than unity.

4. (1) Divergent unless x be less than unity.

(2) Convergent unless x or its modulus be numerically greater than unity.

6. Divergent unless $x < e^{-1}$.

7. x must not be less than unity numerically.

17. See Ex. 18.

CHAPTER IX.

1. (1) $u = \frac{\Delta u}{2x+1}\left(x^2 + \frac{\Delta u}{2x+1}\right)$. (2) The same.

(3) $\left\{\frac{1}{(a-1)^2} - \frac{x}{a-1}\right\}\Delta^2 u_x + x\Delta u_x - u_x = 0.$

(4) $\left(\frac{\Delta u}{a-1}\right)^2 + a^{2x}\left(\frac{\Delta u}{a-1} - u\right) = 0.$ (5) The same.

2. $u_x = Cp^x a^{x(x-1)} + \frac{q}{1-pa}\, a^{(x-1)2}.$

3. $u_x = Ca^x + \frac{\cos(x-1)n - a\cos nx}{1 - 2a\cos n + a^2}.$

4. $u_x = \frac{C + 2^{x+1}}{C + 2^x} - x - 2.$

5. $u_x = \{C + \operatorname{cosec} a \tan(x-1)a\} \cos a \cos 2a \ldots \cos(x-1)a.$

6. Assume $u_x = v_x + m$ where m is a root of

$$m^2 + am + b = 0,$$

and there results a linear equation in $\frac{1}{v_x}$.

8. $u_x = C\epsilon^{x(x-2)} + x\epsilon^{(x-1)^2}.$

9. $u_x = \dfrac{2 \sin x\theta \sin\left(x - \dfrac{1}{2}\right)\theta}{\sin\dfrac{\theta}{2}} + C \sin x\theta.$

10. $u_x = a^{x-1}\{x^2 + C\}.$

11. $u_x = \cos 2^x \theta.$

12. $u_x = a^{x-1}\left\{C + x - \Sigma \dfrac{1}{(x+1)^2}\right\}.$

13. $u_x = \dfrac{1}{2}\{a^{2^x} - a^{-2^x}\}.$

14. $u_x = m^{\frac{n^x - 1}{n-1}} a^{n^x}.$

15. By writing $u_x + \dfrac{1}{2} = v_x$ the equation may be reduced to $v_{x+1} = v_x^2 + C$. When $C = -2$ this gives $v_x = 2 \cos 2^x \theta$.

16. $cu_x = c^2 x + 1.$

17. $\dfrac{u_{x+1}}{u_x} = ax$ or $-2ax$. Hence two associated solutions (see Ch. X.) are

$$u_x = Ca^x \Gamma x$$

and $$u_x = C(-2a)^x \Gamma x.$$

CHAPTER X.

1. $u_x = C + \dfrac{a+2b}{3} x - \dfrac{2(a-b)}{3\sqrt{3}} \cos \pi x \cos \dfrac{\pi}{3}\left(x - \dfrac{1}{2}\right).$

4. $u_x = C^2 + C\dfrac{1-a}{1+a}(-a)^x - \dfrac{a^{2x+1}}{(1+a)^2}.$

8. The two others are given by

$$u_x = \left(cz^x - \frac{z}{z-1}\right)^3,$$

where z is a root of $\mu^2 + \mu + 1 = 0$.

9. $\left(\Delta y - \frac{y}{x} + x + 1\right)\left(\Delta y + \frac{2y}{x-1}\right) = 0,$

$$\Delta y = \left(\frac{y}{x} - x - 1\right)\{1 + (-1)^{x-r}\} - \frac{2y}{x-1}\{1 - (-1)^{x-r}\}.$$

CHAPTER XI.

1. $u_x = C(-1)^x + C'4^x + \dfrac{m^x}{(m+1)(m-4)}.$

2. $u_x = C(-4)^x + \dfrac{5x-6}{25}.$

3. $u_x = (C + C'x)(-1)^x + \dfrac{1}{4}\left[x^3 - 6x^2 + \dfrac{19}{2}x - 3\right] + \dfrac{(-1)^x}{6}x^{(3)}.$

4. $u_x = (m^2 + n^2)^{\frac{x}{2}}\left\{C\cos\left(x\tan^{-1}\dfrac{n}{m}\right) + C'\sin\left(x\tan^{-1}\dfrac{n}{m}\right)\right\} + \dfrac{m^x}{n^2}.$

5. $u_x = C - \dfrac{\cos\left(x - \frac{3}{2}\right)}{2\sin\frac{1}{2}} + \dfrac{(x-1)(x-2)}{2}.$

6. $u_x = (-3)^x\left\{C + \dfrac{1}{192}x\right\} + C_1 + C_2x + C_3x^2 + \dfrac{x^{(3)}}{960}\{4x^2 - 23x + 28\}.$

7. The particular integral is obtained by (II) and (III) page 218. It is any value of $2^{(x-2)}\dfrac{x + \Sigma\cos x}{2E^2 - E - 2}.$

8. $u_x = \dfrac{3^x}{64} - \dfrac{(x+1)^{(7)}}{\lfloor 7} + C\cdot 2^x + C_1 + C_2x + C_3x^2 + C_4x^3 + C_5x^4 + C_6x^5.$

9. $u_x = \dfrac{\pm n^2 \cos mx + \cos (x-2) m}{n^4 \pm 2n^2 \cos 2m + 1}$ + complementary function, which is

$$\left\{C \cos \frac{\pi x}{2} + C' \sin \frac{\pi x}{2}\right\} n^x,$$

or $Cn^x + C'(-n)^x$, according as the upper or lower sign is taken.

10. $$n^{-x} u_x = \{C_1 + C_2 x\} \cos \frac{\pi x}{2} + \{C_3 + C_4 x\} \sin \frac{\pi x}{2},$$

or $$C_1 + C_2 x + \{C_3 + C_4 x\} (-1)^x.$$

11. $(a + bk) \left\{\dfrac{k+1}{k}\right\}^n - bk$ where $k = \dfrac{100m}{r}$.

12. $$\frac{1}{3\sqrt{17}} \left\{\left(\frac{11 + 3\sqrt{17}}{2}\right)^x - \left(\frac{11 - 3\sqrt{17}}{2}\right)^x\right\}.$$

CHAPTER XII.

1. $u_x = \Sigma \Gamma (x-1) \left\{C + \Sigma \dfrac{\sin x}{\Gamma x}\right\} + C'$.

2. $u_x = A \sin (X + \alpha) + B \sin (2X + \beta) + C \sin (3X + \gamma) + \cdots$ to $\dfrac{n-1}{2}$ terms (supposing that n is odd) where $X \equiv \dfrac{2\pi x}{n}$.

3. $u_x = \lfloor x+1 \left\{\Sigma \dfrac{C(-1)^x}{\lfloor x+2} + C'\right\}$.

4. $u_x = \lfloor x \left\{\Sigma \dfrac{C(-1)^x}{\lfloor x+1} + C'\right\}$.

5. $u_x = \lfloor x-3 \left\{C + C'x + \dfrac{x(x-1)(x-2)}{1.2.3}\right\}$.

6. $\log u_x = (-n)^x \left\{C + C'x\right\} + \dfrac{\log a}{(1+n)^2}$.

7. For $(x+2)^2$ read $(x+2)^3$. The equation is then reduced into a very simple form by substituting $\dfrac{x}{(x+1)^2} v_x$ for u_x.

8. $$\left.\begin{aligned} u_x &= C_1 + C_2(-1)^x + mx \\ v_x &= C_1 - C_2(-1)^x - m(x+1) \end{aligned}\right\}.$$

9. $u_x = C_0 + C_1(-1)^x,\ v_x = C_1 - C_0(-1)^x$.

10. $$u_x + \frac{2l-m-n}{3}x = A + \left\{B\cos\frac{\pi x}{3} + C\sin\frac{\pi x}{3}\right\}(-1)^x,$$

and $v_x + \ldots,\ w_x + \ldots$ are obtained by writing $x+2$ and $x+4$ in the quantity on the right-hand side.

11. $$u_x = (A+Bx)2^x + (C+Dx)(-2)^x + \frac{a^{x+2} - 2a^x - 2a^{-x-1}}{(a^2-4)^2},$$

and v_{x+1} (and therefore v_x) is given at once by the first equation.

13. It may be written $(E-a^{-x})(E-a^x)u_x = 0$.

14. $u_x = a^{\frac{x(x-1)}{2}}\{c + C'\Sigma a^{-\frac{x(x-1)}{2}}\}$.

15. $u_x = \sqrt{a}\tan\left\{C_1\cos\dfrac{2\pi x}{3} + C_2\sin\dfrac{2\pi x}{3}\right\}$.

17. Compare with (15) after dividing by $u_x u_{x+1} u_{x+2}$.

19. If $\log a_n = u_n$ we have

$$u_{n+3} + 3u_{n+2} - 4u_n = 0,$$

and the solution is

$$a_n = a\left\{\frac{bc}{a^2}\right\}^{\frac{1-(-2)^n}{3}}$$

20. See page 228. Perform Δ on the equation and a linear equation in $\Delta^2 u_x$ results.

21. $u_x = P\alpha^x + Q\beta^x + R\gamma^x$ where $\alpha\beta\gamma = 1$

and $$C = PQR(\alpha-\beta)^2(\beta-\gamma)^2(\gamma-\alpha)^2.$$

If $C=0$, the solution becomes

$$u_x = P\alpha^x + Q\beta^x.$$

22. If $v_1 = 2\cos\alpha$, $v_m = \dfrac{2\cos\alpha\cos m\alpha - v_0\sin(m-1)\alpha}{\sin\alpha}$.

23. $\dfrac{1}{4}\left\{\left(\dfrac{4}{5}\right)^x - \left(\dfrac{2}{5}\right)^x\right\}$.

CHAPTER XIV.

1. $u_{x,y} = a^x \epsilon^{\frac{y}{a}} \left(\dfrac{d}{dy}\right)^x \phi(y)$.

2. $u_{x,y} = \alpha^x \left(\dfrac{d}{dy}\right)^x \phi_1(y) + \beta^x \left(\dfrac{d}{dy}\right)^x \phi_2(y)$, where α and β are roots of $m^2 - am + b = 0$.

3. $u_{x,y} = x(y+x-1) + \phi(y+x)$.

4. $u_{x,y} = \dfrac{a^{x-y}}{a^{1-n}-1} + \phi(y-nx)$.

5. $u_{x,y} = \dfrac{1}{2}\{a^x C_{y+z} + (-a)^x C'_{y+x}\}$.

6. $f_1(x+y) + xf_2(x+y) + x^2 f_3(x+y) + \dfrac{x^{(4)}}{24}$.

7. $u_x = a^x f_1(x) + b^x f_2(y) + \dfrac{c^{x+y}}{(c-a)(c-b)}$.

8. The complementary function is given by

$$u_x = \alpha^x f_1(x+y) + \beta^x f_2(x+y) + \gamma^x f_3(x+y)$$

where α, β, γ are the roots of

$$m^3 - 3a^2 m + a^3 = 0.$$

The other part is a particular integral of

$$(E^3 - 3a^2 E + a^3)\, u_x = Cx - x^2$$

in which $x+y$ is written for C after solution. It is of the form $Axy + Bx + B'y + C'$, but the values of the coefficients are complicate.

9. $u_x = C2^x + C'\epsilon^x + \dfrac{\epsilon^{nx}}{(\epsilon^n - 2)(n-1)}$ where C is a periodic and C' an absolute constant.

CHAPTER XV.

1. $\phi^n(x) = \sqrt{-1}\, \dfrac{(\sqrt{-1}+x)^m - (\sqrt{-1}-x)^m}{(\sqrt{-1}+x)^m + (\sqrt{-1}-x)^m}$. $(m = 2^n)$.

2. $\phi^n(x) = \dfrac{1}{2}\{(x+\sqrt{x^2-1})^m + (x-\sqrt{x^2-1})^m\}$. $(m = 2^n)$.

5. $f(x) = Cx$.

6. $f(x) = \dfrac{1-(-1)^n a}{1-a^2}\, x^n$.

7. $f(x) = \dfrac{\epsilon^x + a\epsilon^{-x}}{1-a^2}$.

8. $f(x) = f(x) - f(2-x) + \dfrac{2x - x^2}{2}$.

10. $\dfrac{2^{x+1} - 2(-1)^x}{2^{x+1} + (-1)^x}$.

11. $f(x) = \frac{a}{2} + \phi(x) - \phi\left(\frac{1}{x}\right)$.

12. $y = c\epsilon^{\phi(x-a)}$, $\phi(x)$ denoting an odd function of x.

13. Develope $f(\pi)$ in ascending powers of π, and apply the conditions of periodicity.

16. $\phi(x) = \frac{\sin mx}{\sin(mx+c)}$,

$\psi(x) = \frac{\sin c}{\sin(mx+c)}$.

22. $\phi(x) = \{f(x)\}^m$.

CHAPTER XVI.

1. $r = a + f\left(\frac{\theta}{2\pi}\right) - f\left(\frac{\theta - \pi}{2\pi}\right)$ where $f(x)$ satisfies the equation $\Delta f(x) = 0$.

2. Write $\log r$ for r in the answer to the previous question.

INDEX

INDEX

The numbers refer to the pages.

GEOMETRY AND THE IMAGINATION

By D. HILBERT and S. COHN-VOSSEN

Translated from the German by P. NEMENYI.

"A fascinating tour of the 20th century mathematical zoo.... Anyone who would like to see proof of the fact that a sphere with a hole can always be bent (no matter how small the hole), learn the theorems about Klein's bottle—a bottle with no edges, no inside, and no outside—and meet other strange creatures of modern geometry will be delighted with Hilbert and Cohn-Vossen's book."
—*Scientific American.*

"Should provided stimulus and inspiration to every student and teacher of geometry."—*Nature.*

"A mathematical classic. . . . The purpose is to make the reader *see* and *feel* the proofs.... readers can penetrate into higher mathematics with . . . pleasure instead of the usual laborious study."
—*American Scientist.*

"Students, particularly, would benefit very much by reading this book . . . they will experience the sensation of being taken into the friendly confidence of a great mathematician and being shown the real significance of things."—*Science Progress.*

"A person with a minimum of formal training can follow the reasoning.... an important [book]."
—*The Mathematics Teacher.*

—1952. 358 pp. 6x9. 8284-0087-3. **$7.50**

GESAMMELTE ABHANDLUNGEN
(Collected Papers)

By D. HILBERT

Volume I (Number Theory) contains Hilbert's papers on Number Theory, including his long paper on Algebraic Numbers. Volume II (Algebra, Invariant Theory, Geometry) covers not only the topics indicated in the sub-title but also papers on Diophantine Equations. Volume III carries the sub-title: Analysis, Foundation of Mathematics, Physics, and Miscellaneous Papers.

—1932/33/35-66. 1,457 pp. 6x9. 8284-0195-0.

Three vol. set. **$27.50**

PRINCIPLES OF MATHEMATICAL LOGIC

By D. HILBERT and W. ACKERMANN

"As a text the book has become a classic . . . the best introduction for the student who seriously wants to master the technique. Some of the features which give it this status are as follows:

"The first feature is its extraordinary lucidity. A second is the intuitive approach, with the introduction of formalization only after a full discussion of motivation. Again, the argument is rigorous and exact . . . A fourth feature is the emphasis on general extra-formal principles . . . Finally, the work is relatively free from bias . . . All together, the book still bears the stamp of the genius of one of the great mathematicians of modern times."—*Bulletin of the A.M.S.*

—1959. xii + 172 pp. 6x9. 8284-0069-5. **$3.95**

THE DEVELOPMENT OF MATHEMATICS IN CHINA AND JAPAN

By Y. MIKAMI

"Filled with valuable information. Mikami's [account of the mathematicians he knew personally] is an attractive features."

—*Scientific American.*

—1913-62. x + 347 pp. 5⅜x8. [149]

KURVENTHEORIE

By K. MENGER

—1932-63. vi+376 pp. 5⅜x8¼. [172]

GEOMETRIE DER ZAHLEN

By H. MINKOWSKI

—viii + 256 pp. 5½x8¼. [93]

DIOPHANTISCHE APPROXIMATIONEN

By H. MINKOWSKI

—viii + 235 pp. 5¼x8¼. [118]

MORDELL, "Fermat's Last Theorem," see *Klein*

INVERSIVE GEOMETRY

By F. MORLEY and F. V. MORLEY

—xi + 273 pp. 5¼x8¼. [101] **$3.95**

INTRODUCTION TO NUMBER THEORY

By T. NAGELL

A special feature of Nagell's well-known text is the rather extensive treatment of Diophantine equations of second and higher degree. A large number of non-routine problems are given.

—1951-64. Corr. repr. of 1st ed. 309 pp. 5⅜x8. [163]

THE THEORY OF SUBSTITUTIONS

By E. NETTO

Partial Contents: CHAP. I. Symmetric and Alternating Functions. II. Multiple- valued Functions and Groups of Substitutions. III. The Different Values of a Multiple-valued Function and their Algebraic Relation to One Another. IV. Transitivity and Primitivity; Simple and Compound Groups; Isomorphism. V. Algebraic Relations between Functions Belonging to the Same Group . . . VII. Certain Special Classes of Groups. VIII. Analytical Representation of Substitutions. The Linear Group. IX. Equations of Second, Third, Fourth Degrees. Groups of an Equation. X. Cyclotomic Equations. XI. Abelian Equations . . . XIII. Algebraic Solution of Equations. XIV. Group of an Algebraic Equation. XV. Algebraically Solvable Equations.

—In prep. Corr. repr. of 1st ed. 310 pp. 5⅜x8.

THE THEORY OF MATRICES

By F. R. GANTMACHER

This treatise, by one of Russia's leading mathematicians gives, in easily accessible form, a coherent account of matrix theory with a view to applications in mathematics, theoretical physics, statistics, electrical engineering, etc. The individual chapters have been kept as far as possible independent of each other, so that the reader acquainted with the contents of Chapter I can proceed immediately to the chapters that especially interest him. Much of the material has been available until now only in the periodical literature.

Partial Contents. VOL. ONE. I. Matrices and Matrix Operations. II. The Algorithm of Gauss and Applications. III. Linear Operators in an n-Dimensional Vector Space. IV. Characteristic Polynomial and Minimal Polynomial of a Matrix (Generalized Bézout Theorem, Method of Faddeev for Simultaneous Computation of Coefficients of Characteristic Polynomial and Adjoint Matrix, . . .). V. Functions of Matrices (Various Forms of the Definition, Components, Application to Integration of System of Linear Differential Eqns, Stability of Motion, . . .). VI. Equivalent Transformations of Polynomial Matrices; Analytic Theory of Elementary Divisors. VII. The Structure of a Linear Operator in an n-Dimensional Space (Minimal Polynomial, Congruence, Factor Space, Jordan Form, Krylov's Method of Transforming Secular Eqn, . . .). VIII. Matrix Equations (Matrix Polynomial Eqns, Roots and Logarithm of Matrices, . . .). IX. Linear Operators in a Unitary Space. X. Quadratic and Hermitian Forms.

VOL. TWO. XI. Complex Symmetric, Skew-symmetric, and Orthogonal Matrices. XII. Singular Pencils of Matrices. XIII. Matrices with Non-Negative Elements (Gen'l and Spectral Properties, Reducible M's, Primitive and Imprimitive M's, Stochastic M's, Totally Non-Negative M's, . . .). XIV. Applications of the Theory of Matrices to the Investigation of Systems of Linear Differential Equations. XV. The Problem of Routh-Hurwitz and Related Questions (Routh's Algorithm, Lyapunov's Theorem, Infinite Hankel M's, Supplements to Routh-Hurwitz Theorem, Stability Criterion of Liénard and Chipart, Hurwitz Polynomials, Stieltjes' Theorem, Domain of Stability, Markov Parameters, Problem of Moments, Markov and Chebyshev Theorems, Generalized Routh-Hurwitz Problem, . . .). BIBLIOGRAPHY.

—Vol. I. 1960. x + 374 pp. 6x9. 8284-0131-4. **$7.50**
—Vol. II. 1960. x + 277 pp. 6x9. 8284-0133-0. **$6.50**

UNTERSUCHUNGEN UEBER HOEHERE ARITHMETIK

By C. F. GAUSS

In this volume are included all of Gauss's number-theoretic works: his masterpiece, *Disquisitiones Arithmeticae*, published when Gauss was only 25 years old; several papers published during the ensuing 31 years; and papers taken from material found in Gauss's handwriting after his death.

These papers (pages 457-695 of the present book) include a fourth, fifth, and sixth proof of the Quadratic Reciprocity Law, researches on biquadratic residues, quadratic forms, and other topics.

—1889-65. xv + 695 pp. 6x9. 8284-0191-8. **$8.75**

THEORY OF PROBABILITY

By B. V. GNEDENKO

This textbook, by Russia's leading probabilist, is suitable for senior undergraduate and first-year graduate courses. It covers, in highly readable form, a wide range of topics and, by carefully selected exercises and examples, keeps the reader throughout in close touch with problems in science and engineering.

The translation has been made from the fourth Russian edition by Prof. B. D. Seckler. Earlier editions have won wide and enthusiastic acceptance as a text at many leading colleges and universities.

"extremely well written . . . suitable for individual study . . . Gnedenko's book is a milestone in the writing on probability theory."—*Science.*

Partial Contents: I. The Concept of Probability (Various approaches to the definition. Space of Elementary Events. Classical Definition. Geometrical Probability. Relative Frequency. Axiomatic construction . . .). II. Sequences of Independent Trials. III Markov Chains IV. Random Variables and Distribution Functions (Continuous and discrete distributions. Multidimensional d. functions. Functions of random variables. Stieltjes integral). V. Numerical Characteristics of Random Variables (Mathematical expectation. Variance...Moments). VI. Law of Large Numbers (Mass phenomena. Tchebychev's form of law. Strong law of large numbers...). VII. Characteristic Functions (Properties. Inversion formula and uniqueness theorem. Helly's theorems. Limit theorems. Char. functs. for multidimensional random variables...). VIII. Classical Limit Theorem (Liapunov's theorem. Local limit theorem). IX. Theory of Infinitely Divisible Distribution Laws. X. Theory of Stochastic Processes (Generalized Markov equation. Continuous S. processes. Purely discontinuous S. processes. Kolmogorov-Feller equations. Homogeneous S. processes with independent increments. Stationary S. process. Stochastic integral. Spectral theorem of S. processes. Birkhoff-Khinchine ergodic theorem). XI. Elements of Queueing Theory (General characterization of the problems. Birth-and-death processes. Single-server queueing systems. Flows. Elements of the theory of stand-by systems). XII. Elements of Statistics (Problems. Variational series. Glivenko's Theorem and Kolmogorov's criterion. Two-sample problem. Critical region . . . Confidence limits). TABLES. BIBLIOGRAPHY. ANSWERS TO THE EXERCISES.

—4th ed. 1968. 527 pp. 6x9. 8284-0132-2. **$9.50**

A SHORT HISTORY OF GREEK MATHEMATICS

By J. GOW

A standard work on the history of Greek mathematics, with special emphasis on the Alexandrian school of mathematics.

—1884-68. xii + 325 pp. 5⅜x8. 8284-0218-3. **$6.50**

SQUARING THE CIRCLE, and other Monographs

By HOBSON, HUDSON, SINGH, and KEMPE

FOUR VOLUMES IN ONE.

SQUARING THE CIRCLE, by *Hobson.* A fascinating account of one of the three famous problems of antiquity, its significance, its history, the mathematical work it inspired in modern times, and its eventual solution in the closing years of the last century.

RULER AND COMPASSES, by *Hudson.* "An analytical and geometrical investigation of how far Euclidean constructions can take us. It is as thoroughgoing as it is constructive."—*Sci. Monthly.*

THE THEORY AND CONSTRUCTION OF NON-DIFFERENTIABLE FUNCTIONS, by *Singh.* I. Functions Defined by Series. II. Functions Defined Geometrically. III. Functions Defined Arithmetically. IV. Properties of Non-Differentiable Functions.

HOW TO DRAW A STRAIGHT LINE, by *Kempe.* An intriguing monograph on linkages. Describes, among other things, a linkage that will trisect any angle.

"Intriguing, meaty."—*Scientific American.*

—388 pp. 4½x7½. 8284-0095-4. Four vols. in one. **$4.95**

SPHERICAL AND ELLIPSOIDAL HARMONICS

By E. W. HOBSON

"A comprehensive treatise . . . and the standard reference in its field."—*Bulletin of the A. M. S.*

—1931-65. xi + 500 pp. 5⅜x8. 8284-0104-7. **$7.50**

ELASTOKINETIK: Die Methoden zur Angenäherten Lösung von Eigenwertproblemen in der Elastokinetik

By K. HOHENEMSER

—(Erg. der Math.) 1932-49. 89 pp. 5½x8½. 8284-0055-5. **$2.75**

ERGODENTHEORIE, *by E. HOPF.* See BEHNKE

RULER AND COMPASSES, *by H. P. HUDSON.* See HOBSON

PHYSIKALISCH-MATHEMATISCHE MONOGRAPHIEN

By W. v. IGNATOWSKY, et al.

THREE VOLUMES IN ONE.

Of the many well-known monographs in the series published by the Steklov Institute of the Academy of Sciences of the U.S.S.R., only a few were originally published in a language other than Russian. Two of the French-language works have been reprinted and are listed elsewhere in this catalogue: Gunther's book on *Stieltjes Integrals* and Lappo-Danilevski's three-volume work on *Systems of Differential Equations.*

CONTENTS: 1. *Untersuchungen einiger Integrale mit Besselschen Funktionen und ihre Anwendung auf Beugungserscheinungen,* by Ignatowsky. 2. *Kreisscheibenkondensator,* by Ignatowsky. 3. *Table of a Special Function,* by Bursian and Fock.

—1932-66. 16 + 232 pp. 6¼x9¼. 8284-0201-9. Three vols. in one. **$5.50**

THE MATHEMATICAL THEORY OF THE TOP,
by F. KLEIN. See SIERPINSKI

FAMOUS PROBLEMS, and other monographs
By KLEIN, SHEPPARD, MacMAHON, and MORDELL

FOUR VOLUMES IN ONE.

FAMOUS PROBLEMS OF ELEMENTARY GEOMETRY, by *Klein*. A fascinating little book. A simple, easily understandable, account of the famous problems of Geometry—The Duplication of the Cube, Trisection of the Angle, Squaring of the Circle—and the proofs that these cannot be solved by ruler and compass—presentable, say, before an undergraduate math club (no calculus required). Also, the modern problems about transcendental numbers, the existence of such numbers, and proofs of the transcendence of *e*.

FROM DETERMINANT TO TENSOR, by *Sheppard*. A novel and charming introduction. Written with the utmost simplicity. PT I. Origin of Determinants. II. Properties of Determinants. III. Solution of Simultaneous Equations. IV. Properties. V. Tensor Notation. PT II. VI. Sets. VII. Cogredience, etc. VIII. Examples from Statistics. IX. Tensors in Theory of Relativity.

INTRODUCTION TO COMBINATORY ANALYSIS, by *MacMahon*. A concise introduction to this field. Written as introduction to the author's two-volume work.

THREE LECTURES ON FERMAT'S LAST THEOREM, by *Mordell*. This famous problem is so easy that a high-school student might not unreasonably hope to solve it; it is so difficult that tens of thousands of amateur and professional mathematicians, Euler and Gauss among them, have failed to find a complete solution. Mordell's very small book begins with an impartial investigation of whether Fermat himself had a solution (as he said he did) and explains what has been accomplished. This is one of the masterpieces of mathematical exposition.

—2nd ed. 1962. 350 pp. 5⅜x8. Four vols. in one.
8284-0108-X. Cloth **$4.95**
8284-0166-7. Paper **$1.95**

VORLESUNGEN UEBER NICHT-EUKLIDISCHE GEOMETRIE
By F. KLEIN

—1928-59. xii + 326 pp. 5x8. 8284-0129-2. **$6.00**

ENTWICKLUNG DER MATHEMATIK IM 19. JAHRHUNDERT
By F. KLEIN

TWO VOLUMES IN ONE.

Vol. I treats of the various branches of advanced mathematics of the prolific 19th century; Klein himself was in the forefront of the mathematical activity of latter part of the 19th and early part of the 20th centuries.

Vol. II deals with the mathematics of relativity theory.

—1926/27-67. 616 pp. 5¼x8. 8284-0074-1. Two vols. in one.
$8.95